New Waves in Electromagnetic Technology

Other related titles:

You may also like

• SBEW533 | Ergul | New Trends in Computational Electromagnetics | 2019

We also publish a wide range of books on the following topics:
Computing and Networks
Control, Robotics and Sensors
Electrical Regulations
Electromagnetics and Radar
Energy Engineering
Healthcare Technologies
History and Management of Technology
IET Codes and Guidance
Materials, Circuits and Devices
Model Forms
Nanomaterials and Nanotechnologies
Optics, Photonics and Lasers
Production, Design and Manufacturing
Security
Telecommunications
Transportation

All books are available in print via https://shop.theiet.org or as eBooks via our Digital Library https://digital-library.theiet.org.

SciTech ELECTROMAGNETIC WAVES 557

Other volumes in this series:

Volume 1 Geometrical Theory of Diffraction For Electromagnetic Waves, 3rd Edition G.L. James
Volume 10 Aperture Antennas and Diffraction Theory E.V. Jull
Volume 11 Adaptive Array Principles J.E. Hudson
Volume 12 Microstrip Antenna Theory and Design J.R. James, P.S. Hall and C. Wood
Volume 15 The Handbook of Antenna Design, Volume 1 A.W. Rudge, K. Milne, A.D. Oliver and P. Knight (Editors)
Volume 16 The Handbook of Antenna Design, Volume 2 A.W. Rudge, K. Milne, A.D. Oliver and P. Knight (Editors)
Volume 18 Corrugated Horns for Microwave Antennas P.J.B. Clarricoats and A.D. Oliver
Volume 19 Microwave Antenna Theory and Design S. Silver (Editor)
Volume 21 Waveguide Handbook N. Marcuvitz
Volume 23 Ferrites at Microwave Frequencies A.J. Baden Fuller
Volume 24 Propagation of Short Radio Waves D.E. Kerr (Editor)
Volume 25 Principles of Microwave Circuits C.G. Montgomery, R.H. Dicke and E.M. Purcell (Editors)
Volume 26 Spherical Near-Field Antenna Measurements J.E. Hansen (Editor)
Volume 28 Handbook of Microstrip Antennas, 2 Volumes J.R. James and P.S. Hall (Editors)
Volume 31 Ionospheric Radio K. Davies
Volume 32 Electromagnetic Waveguides: Theory and applications S.F. Mahmoud
Volume 33 Radio Direction Finding and Superresolution, 2nd Edition P.J.D. Gething
Volume 34 Electrodynamic Theory of Superconductors S.A. Zhou
Volume 35 VHF and UHF Antennas R.A. Burberry
Volume 36 Propagation, Scattering and Diffraction of Electromagnetic Waves A.S. Ilyinski, G.Ya. Slepyan and A.Ya. Slepyan
Volume 37 Geometrical Theory of Diffraction V.A. Borovikov and B.Ye. Kinber
Volume 38 Analysis of Metallic Antenna and Scatterers B.D. Popovic and B. M. Kolundzija
Volume 39 Microwave Horns and Feeds A.D. Olver, P.J.B. Clarricoats, A.A. Kishk and L. Shafai
Volume 41 Approximate Boundary Conditions in Electromagnetics T.B.A. Senior and J.L. Volakis
Volume 42 Spectral Theory and Excitation of Open Structures V.P. Shestopalov and Y. Shestopalov
Volume 43 Open Electromagnetic Waveguides T. Rozzi and M. Mongiardo

New Waves in Electromagnetic Technology

Edited by
Andrew Michael Chugg

The Institution of Engineering and Technology

About the IET

This book is published by the Institution of Engineering and Technology (The IET).

We inspire, inform and influence the global engineering community to engineer a better world. As a diverse home across engineering and technology, we share knowledge that helps make better sense of the world, to accelerate innovation and solve the global challenges that matter.

The IET is a not-for-profit organisation. The surplus we make from our books is used to support activities and products for the engineering community and promote the positive role of science, engineering and technology in the world. This includes education resources and outreach, scholarships and awards, events and courses, publications, professional development and mentoring, and advocacy to governments.

To discover more about the IET please visit https://www.theiet.org/.

About IET books

The IET publishes books across many engineering and technology disciplines. Our authors and editors offer fresh perspectives from universities and industry. Within our subject areas, we have several book series steered by editorial boards made up of leading subject experts.

We peer review each book at the proposal stage to ensure the quality and relevance of our publications.

Get involved

If you are interested in becoming an author, editor, series advisor, or peer reviewer please visit https://www.theiet.org/publishing/publishing-with-iet-books/ or contact author_support@theiet.org.

Discovering our electronic content

All of our books are available online via the IET's Digital Library. Our Digital Library is the home of technical documents, eBooks, conference publications, real-life case studies and journal articles. To find out more, please visit https://digital-library.theiet.org.

In collaboration with the United Nations and the International Publishers Association, the IET is a Signatory member of the SDG Publishers Compact. The Compact aims to accelerate progress to achieve the Sustainable Development Goals (SDGs) by 2030. Signatories aspire to develop sustainable practices and act as champions of the SDGs during the Decade of Action (2020–30), publishing books and journals that will help inform, develop, and inspire action in that direction.

In line with our sustainable goals, our UK printing partner has FSC accreditation, which is reducing our environmental impact to the planet. We use a print-on-demand model to further reduce our carbon footprint.

British Library Cataloguing in Publication Data
A catalogue record for this product is available from the British Library

ISBN 978-1-83953-456-0 (hardback)
ISBN 978-1-83953-457-7 (PDF)

Typeset in India by MPS Limited

Cover image: Fantastic Connection : Foto-Ruhrgebiet/istock via Getty images

Contents

Preface
Leading the charge in making waves

Andrew Michael Chugg[1]

The concept for this book emerged out of a webinar on "Electromagnetic Waves: Successfully Surfing the Subject" that I gave on behalf of the IET at the height of the COVID crisis on Tuesday 16 June 2020. The event proved successful both in terms of the number of attendees and the enthusiasm of the audience for the subject matter. Perhaps this is attributable to the broad appeal of a subject that literally casts light upon the whole field of science or else perhaps it was facilitated by a thirst for distractions from the stresses of lock-down. At any rate, the happy outcome was an invitation by an agent of the publishers to propose a book on related subject areas.

In response, my proposal was a volume dedicated to the theme of the future of electromagnetic technology. The concept was that experts in specialist areas of electromagnetics would write each chapter and provide insightful summaries of current developments within their specialisms. They would then extrapolate these existing trends in a logical and reasoned fashion to forecast future technological developments in their field. The challenge was for them to look ahead by anything up to a century or more.

This volume presents the fruits of this plan in focusing on current and forthcoming developments in ten of the most exciting and important areas of technological development involving the field of electromagnetics. Specifically, the first chapter addresses wireless power transfer (WPT) across small gaps and looks at the future spread of this technology to an ever-broader range of applications, systems and vehicles, even including aircraft and cars whilst in motion. WPT antennas are predicted to scale up to run transport systems, but also to scale down to supply medical implants. It also links the future of WPT to advances in battery technology, such as higher density power storage in batteries.

Second, the state of progress in magnetic confinement fusion is reviewed, and it is shown that this technology is progressing rapidly towards delivering practical commercial systems to the market, potentially as soon as 2050. Third, the chapter on the role of metamaterials examines their applications to improve antennas, lenses and cloaking technologies such as stealth and how these capabilities are rapidly evolving. This leads to a predicted linkage between the shrinkage of metamaterial cell dimensions and the feasibility of cloaking devices. Fourth, there is a chapter on future developments in superconducting motors and flux pumps.

---

[1]Executive Technical Expert in Electromagnetics and Radiation Effects at MBDA UK Limited

This is an analysis driven by practical commercial considerations of the applications for these devices and their pros and cons relative to conventional motors. The advantages and drawbacks of high-temperature superconductors versus low-temperature superconductors are also examined, leading to specific predictions of the technology areas for early commercial exploitation of superconducting motors.

Magnetic levitation technologies and their applications are reviewed in chapter 5, including uses ranging from magnetic levitation trains to virtual reality simulations of surgical procedures. There is an interesting prediction that magnetic levitation vehicles will require evacuated pathways to achieve a decisive technological efficiency edge over rival systems, either in the form of vacuum tunnels or on airless bodies such as the Moon.

The sixth chapter focuses on dramatic healthcare improvements that are forecast to be achieved through better control of electrostatic buildups in healthcare facilities using dissipative equipment, ESD-proofing of personal protective equipment (PPE) and improved facility design. The benefits will arise in a variety of ways, especially reduced attraction of pathogens onto medical equipment by electrostatic fields.

In a seventh chapter, the question is posed of whether electromagnetic propulsion systems might theoretically be capable of taking a spacecraft carrying a human crew to the nearby stars on a timescale shorter than their lifespans? There is an analysis of a possible approach to achieving this, which shows that the engineering challenges are immense and unlikely to be met in the near future, Nevertheless, similar electromagnetic propulsion is likely to be involved in interplanetary transportation within the next century.

The eighth chapter in the field of photonics focusses specifically on liquid crystal displays (LCDs) and the associated technology. Whereas LCD screens have become the dominant display technology in recent decades, this review points out that LCDs are developing applications far beyond simple screen displays, including the reconfiguration of light paths through instrumentation and optical information processing.

Novel and advanced particle accelerators are not only important for high-energy physics experiments but also have key technological applications, such as tomographic scanners for three-dimensional imaging within structures or even for medical scans on people. In the ninth chapter, there is a forecast and description of technology capable of accelerating particles to ever higher energies in narrower beams within reduced volumes, implying great improvements in these tomographic scanners.

Finally, peering a little further into the technological future, there is a discussion of the manner in which magnetic monopoles seem to be required to complete the electromagnetic theory yet have proved supremely elusive in the face of an extensive and multi-faceted hunt to track them down. This leads to an analysis of the various ways in which the availability of such particles might influence future technologies, albeit on a timescale of centuries.

It is an important consequence of the Maxwell equations, which constitute the theoretical framework of electromagnetics, that accelerating or oscillating a charge generates a wave in the electromagnetic field. In a parallel manner, this book aims to lead the charge in making waves in the field of electromagnetic technology.

About the editor

Andrew Michael Chugg read Natural Sciences at Trinity College in the University of Cambridge in the UK, graduating with honours. Since 1997, he has been the author of over 20 papers on radiation effects, mostly published in *IEEE Transactions on Nuclear Science*, and in 2013, he was technical chair of the RADECS Conference held in Oxford. He is currently an executive technical expert in Electromagnetics & Radiation Physics at MBDA in Bristol, UK.

Chapter 1
Wireless power transfer across small gaps

Jiafeng Zhou[1] and Jinyao Zhang[1]

Wireless power transfer (WPT) technologies can be categorized into two types: near-field WPT (based on coupling) and far-field WPT (based on radiation). Near-field WPT can be further divided into inductive and capacitive types, depending on how energy is transferred. The first major breakthrough in WPT technology was made by Nikola Tesla, who demonstrated wireless lighting for phosphorescent lamps using inductive coupling with a Tesla coil. Near-field inductive WPT involves energy transfer over short distances (typically tens of mm) through aligned coils and inductive coupling. Capacitive coupling WPT uses electric field coupling to transfer energy between closely spaced metal plates. Today, many consumer electronics, such as mobile phones, electric toothbrushes and Bluetooth headsets, can be wirelessly charged over short distances using charging mats or cases. WPT technologies are also used in higher-power applications, such as charging computers and even vehicles. Far-field WPT uses microwave radiation for energy transfer over medium to long distances. The greater transfer distance allows more flexibility in the placement of the transmitter and receiver. Far-field WPT can power low-power devices like wearables, implants and Internet of Things (IoT) sensors, as well as medium to high-power devices like smartphones, tablets and drones.

1.1 Inductive coupling WPT

1.1.1 Magnetic induction coupling

Inductive coupling WPT refers to the transfer of electrical energy between a power source and a load using the principle of electromagnetic induction across a small gap. This method relies on the transfer of energy through a magnetic field that changes over time. As illustrated in Figure 1.1, energy is transferred to the secondary side due to the magnetic flux produced by the current I_1 flowing in the primary side and passing through a secondary-side coil. This results in a voltage V being induced in the opposite direction to the magnetic flux on the secondary side, allowing power to be transmitted in the form of current flow I_2. Energy transfer occurs through changes in the magnetic

[1]Department of Electrical Engineering and Electronics, University of Liverpool, UK

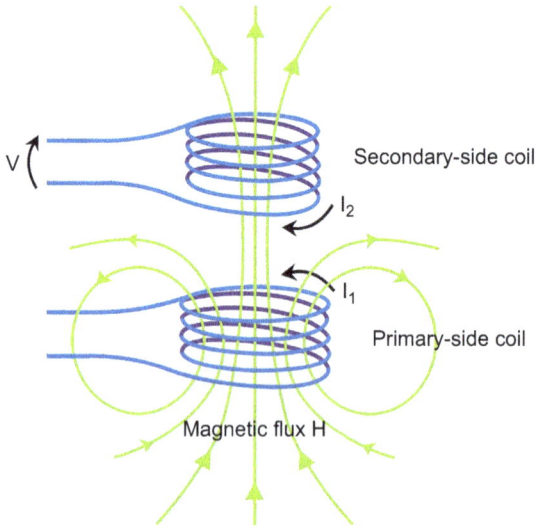

Figure 1.1   Power transfer by electromagnetic induction

field over time dH/dt. The power source generates an alternating current in the primary coil, which creates a magnetic field that induces a current in the secondary coil. This current is then rectified to provide direct current (DC) power to the load.

1.1.2   Resonant inductive coupling

The difference between magnetic induction and magnetic resonance coupling lies in the utilization of resonance. Figure 1.2 shows four types of circuit topologies: N–N (no resonance), N–S (secondary-side resonance), S–N (primary-side resonance) and S–S (resonance on both primary- and secondary-side, which is also called magnetic resonant coupling). The letter 'S' here indicates that the resonant capacitor is connected in series with the coil to realize resonance. Any of the series capacitors can be replaced by one that is connected to the coil in parallel. 'P' can be used to represent these cases. There will be nine types of connections in total. The other five types are N–P, P–N, S–P, P–S and P–P.

Figures 1.3 and 1.4 depict the input power and power consumed by the load for four types of circuits with a large air gap (weak coupling), along with their corresponding calculated efficiency, which is defined as the received power by the load resistor divided by the source power [1]. The calculation was conducted using a constant-voltage source, with all test conditions kept the same across the four circuit types, including coil size, number of coil turn and distance between coils. The only variable was whether resonant capacitors were added.

Table 1.1 summarizes the efficiency and received power of the four circuit types mentioned above, based on a constant-voltage source. When a resonant capacitor C_1 is inserted on the primary side only, the power received by the secondary coil increases with its insertion. However, the efficiency remains nearly

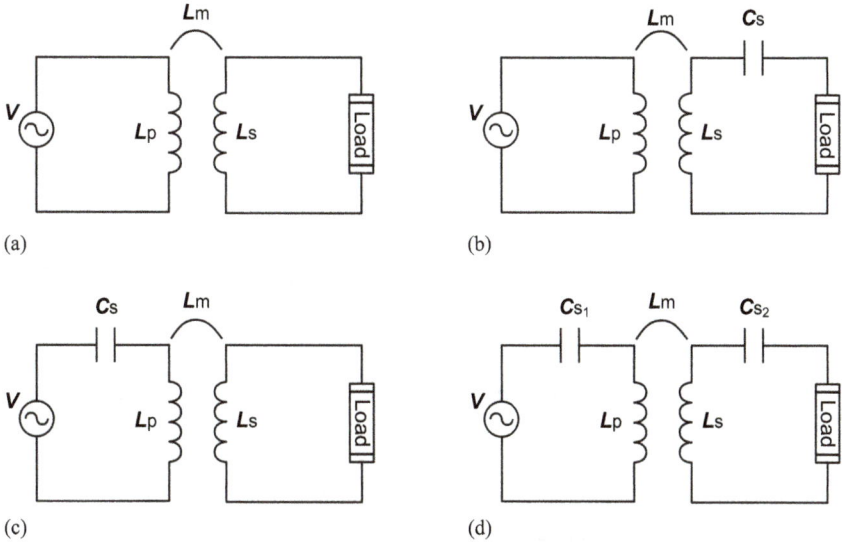

Figure 1.2 Mono-resonant compensation topologies. (a) Nonresonance (N–N). (b) Secondary-side resonance (N–S). (c) Primary-side resonance (S–N). (d) Magnetic resonant coupling (S–S).

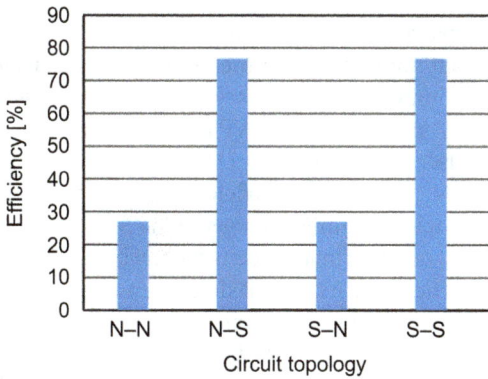

Figure 1.3 Comparison of efficiency of four types of circuits [1]

constant, because of the exponential increase in power sent by the primary coil. Conversely, in a circuit with only a resonant capacitor C_2 inserted on the secondary side, the efficiency increases with its insertion, but the power received by the secondary coil increases only slightly. This is because the power sent by the primary coil is initially quite low, and the resonance of the secondary coil does not cause an increase in the power sent by the primary coil, so the power increase resulting from the increase in efficiency is limited. It can be seen that if only one

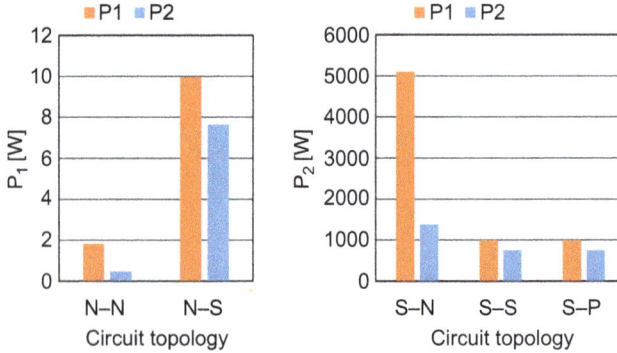

Figure 1.4   Comparison of P_1 (input power) and P_2 (power consumed by load) of the four types of circuits [1]

Table 1.1   Efficiency η and power P_2 of four types of circuit [1]

	η	P_2
N–N	Low	Low
N–S	High	Low
S–N	Low	High
S–S	High	High

side of the circuit is inserted with a resonant capacitor, the increase is only either power or efficiency, but not both. Therefore, a magnetic resonance coupled circuit with two series capacitors (on the primary and secondary sides) is required to achieve high efficiency and high power WPT over a relatively large air gap.

As an example case, it was shown in [1] that the flux distribution of the N–N circuit is similar to the N–S circuit, but the field can be more than ten times stronger in the S–S circuit or more than 100 times stronger in the S–N case. In the N–N and S–N circuits, the flux density on the primary side is high, while on the secondary side, it is low, resulting in low efficiency. As the maximum magnetic flux strength is normalized, although their flux distributions are similar, the latter has a higher flux density due to resonance. In the N–S and S–S circuits, the primary and secondary side coils have nearly equal magnetic flux densities, leading to high efficiency. However, the flux density in the N–S is very low, while it is very high in the S–S circuit. As a result, the S–S circuit exhibits high efficiency and can receive high power.

1.1.3   Wireless transformer pads

In magnetic coupling WPT systems, the transmitter and receiver pads are constructed from multiple component layers to achieve maximum power transfer efficiency and minimize electromagnetic interference. As depicted in Figure 1.5, the main components of a wireless transformer pad are the coil, the shielding material (ferrite and aluminium plate) and the protective and supportive layers [2].

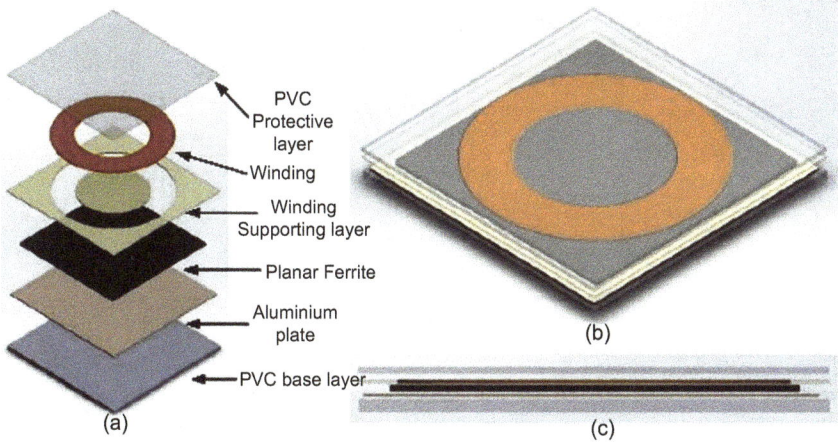

Figure 1.5   The exploded view of the wireless transformer pads [2]

The shape of the transmitter and receiver coils can greatly influence the efficiency of energy transfer. Circular and rectangular coils are very common, while some hybrid coil configurations have been employed in designs to enhance performance and accommodate misalignment between transmitter and receiver pads. The magnetic ferrite structure is another essential component of the pad. It reduces the coupling effectiveness between two coils, especially if there is no shielding to reduce leakage fluxes. Magnetic ferrite cores can redirect the magnetic flux from the primary coil to the secondary coil and improve the mutual inductance and self-inductance of the coils. During wireless transformer pad design, several factors such as the shape, size, permeability and operating frequency of the ferrite core need to be taken into consideration.

1.2   Capacitive coupling WPT

Power transfer can be achieved not only by magnetic field coupling, but also by electric field coupling. This method requires closer distances compared to inductive coupling ones, as the energy transfer relies on close proximity between the metal plates. A schematic of the electric field coupling structure is shown in Figure 1.6. It consists of four metal plates, two on the transmitter side and two on the receiver side. The metal plates are typically separated by an insulating material, such as air or plastic. The transmitter side is connected to the power supply to create an electromagnetic field between the conductive plates, while the receiver side is connected to the load. Energy is transmitted due to the coupling of the electric field generated on the primary side plate with the secondary side plate. Similar to the inductive coupling case, the propagation of energy is caused by the electric field changing over time. Capacitive coupling WPT is often used to power some medical applications, such as pacemakers, hearing aids and electric vehicles (EVs).

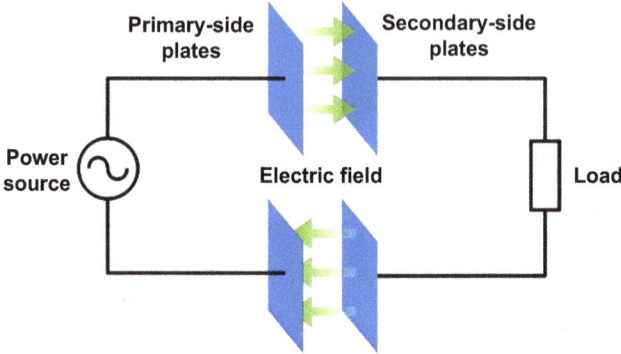

Figure 1.6   Schematic diagram of capacitive coupling

The types of electric coupling can also be classified as non-resonant electric field coupling (N–N), primary-side resonance (S–N), secondary-side resonance (N–S) and double-side resonance (S–S) based on the presence or absence of resonance. Similar to magnetic resonant coupling, electric resonant coupling (S–S) can provide both high energy efficiency and high received power. The difference is that resonance is achieved by adding inductors, rather than capacitors as used in magnetic resonance coupling circuits.

1.3   Microwave power transfer

Microwave power transfer is a far-field WPT technology based on the transfer of electrical energy from a power source to a load device using electromagnetic waves in the form of radiation. Unlike coupling-based WPT, radiation-based WPT transmits energy over greater distances and can send it out even if there is no receiver nearby. The radio frequency (RF) power emitted can reach micro-watt to milli-watt levels at the receiver end, which can be used for the battery charging of low-power electronics or for powering battery-free devices.

Microwave power transfer (also known as far-field WPT) is a promising technology that has a wide range of applications in various fields, including medical applications, such as powering implantable devices, or the charging of wearable medical devices. It can also be used for industrial or agricultural applications, such as powering a large range of remote sensors, eliminating the need for battery replacement and reducing maintenance costs. The use of wirelessly powered small industrial robots or unmanned aerial vehicles (drones) can reduce the need for physical connections and increase their mobility and operational efficiency. There are many potential applications of far-field WPT. As the technology continues to advance, new innovative uses are likely to emerge, for example, wireless charging of electronic devices such as smartphones, laptops and wearables in a defined space.

The basic structure of a far-field WPT system is shown in Figure 1.7. The RF signal generated by an RF source is boosted by a power amplifier, then broadcast

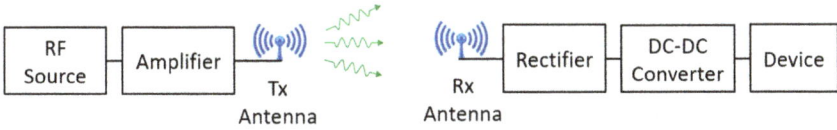

Figure 1.7   Basic structure of a far-field WPT system

over the air through one or more transmit antennas and finally collected and converted to DC in the RF receiver. The receiver consists of one or more receiving antennas with one or more rectifiers (the combination of an antenna and a rectifier is called a rectenna). The rectifier converts the RF power received by the receiving antenna into DC power, which can then be used to power the load device. The design of the RF–DC conversion circuit can be broadly classified into Schottky diode rectifiers and synchronous rectifiers based on high-electron-mobility transistors. RF–DC conversion circuits based on Schottky diodes typically have a low knee voltage (forward voltage drop) and provide a very good RF to DC conversion efficiency at a low power level, but their low reverse voltage rating makes them unsuitable for high-power applications. Conversely, synchronous rectifiers can achieve high efficiency at high input power due to the higher breakdown voltage of high-electron-mobility transistors. As the output DC voltage of the rectifier can vary considerably depending on the RF input power, a DC–DC converter or power management unit is needed to convert the DC voltage to a suitable value for the electronic device intended as the load.

The end-to-end power transfer efficiency in a far-field WPT system is the ratio of the DC power successfully received by the load at the receiving end to the RF power at the transmitting end. It can be defined as

$$\eta = \frac{P_{dc}^r}{P_{dc}^t} = \underbrace{\frac{P_{rf}^t}{P_{dc}^t}}_{\eta_1} \underbrace{\frac{P_{rf}^r}{P_{rf}^t}}_{\eta_2} \underbrace{\frac{P_{dc}^r}{P_{rf}^r}}_{\eta_3}$$

where η_1, $/\eta_2$ and η_3 denote the DC-to-RF, RF-to-RF and RF-to-DC power conversion efficiencies, respectively. The performance at the transmitting end (DC-to-RF) depends primarily on the efficiency of the RF power amplifier and the transmitting antenna. The design of the transmitting and receiving antennas plays a crucial role in power transfer efficiency. For example, the use of high-gain directional and array antennas can increase the efficiency of power transfer by focusing electromagnetic waves on the receiving end. The distance between the transmitting and receiving antennas affects the RF power transfer efficiency (RF-to-RF), which decreases as the distance increases. The frequency used in the far-field WPT system also affects the power transfer efficiency, with higher frequencies typically having higher free-space path loss. This includes interference from other sources in space and the effects of environmental conditions such as atmospheric absorption and reflection, which can reduce power transfer efficiency. The factors that determine the efficiency of the receiver side are mainly the receiver antenna design and the RF rectifier (RF-to-DC). The main challenges for far-

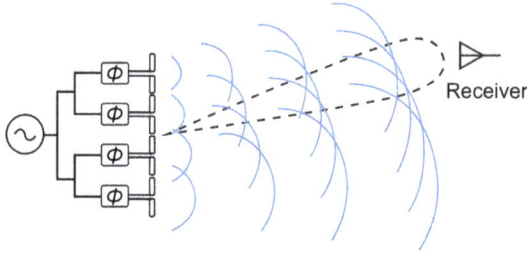

Figure 1.8   Diagram of phased array antennas

field WPT are the low level of RF power received by receiving antennas (relative to near-field WPT) and the low efficiency of the RF-DC power conversion.

Microwave power transfer systems typically operate in the frequency range of 300 MHz to 300 GHz. When receiver antennas and rectifiers that are individually designed are combined, impedance matching is required to reduce reflection losses and thus improve overall system efficiency. However, impedance matching circuits can be complex, which may not be feasible for applications that are size and weight-sensitive. The advent of the rectenna has addressed this issue by directly matching the output impedance of the receiving antenna to the input impedance of the rectifier. By integrating the receiving antenna and rectifier in this manner, the reflection and transfer losses can be reduced without the need for additional matching networks.

Phased array antennas are able to focus the emitted energy in a specific and adjustable direction, thus reducing the energy loss as it propagates from the transmitting antenna to the receiving antenna. In a phased array system, the energy from the transmitter is fed to the radiating components typically through phase shifters, which are controlled by a computing system that can electronically adjust the phase or signal delay, directing the beam waves towards a different direction, as Figure 1.8 shows. The beam direction can be precisely controlled by adjusting the phase delay of each antenna in the array. This allows the beam direction, i.e. the direction of energy transfer, to be dynamically adjusted according to changes in the position of the receiver. Phased array antenna systems are often used in WPT applications that require high power transfer over long distances, such as powering portable devices or drones. The high gain of phased array antennas allows WPT systems based on them to transmit energy over longer distances with less energy loss.

1.4   WPT for portable device charging

Inductive-coupling-based WPT is already extensively used in charging small devices such as smartphones and electric toothbrushes. The primary coil is typically incorporated into the charging base, while the secondary coil is integrated into the device to be charged. The close proximity between the coils enables efficient energy transfer. Figure 1.9 shows a basic WPT system for mobile phone charging, where the secondary coil is integrated into the portable device, and the primary coil is located on the charging pad.

Screen Assembly
Circuit Board

Battery

Induction Coil

Wireless Charger

Induction Coil

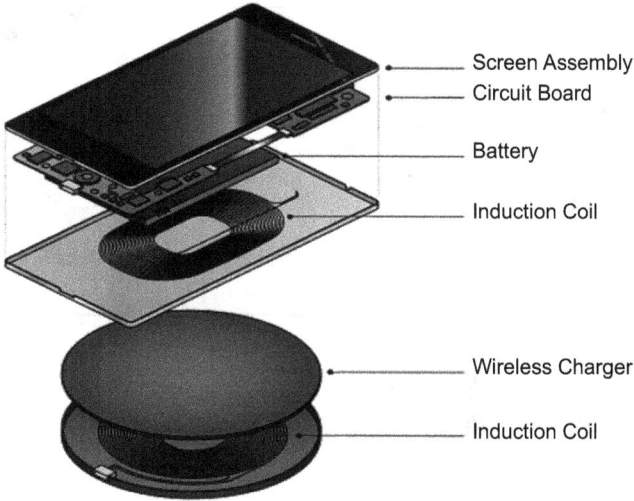

Figure 1.9   The exploded view of a mobile phone WPT system [3]

Charging pads for WPT of mobile devices usually consist of a flat surface on which the mobile device can be placed for charging. The design of the coils in a charging pad is crucial for its effectiveness and performance. Several factors, such as the number of coils, coil shape, size, number of turns and materials used, must be considered to optimize the power transfer and provide a convenient charging experience for the user.

1.4.1   Alignment methods

The alignment of the coils in the charging system plays an important role in the power transfer efficiency and charging distance. The charging pad can have a single coil or multiple coils, each of which can provide different alignments. While a single coil can offer power transfer over a limited distance, multiple coils can expand the charging area and offer greater placement flexibility for the mobile device. Three different alignment methods are shown in Figure 1.10 [4].

As illustrated in Figure 1.10(a), magnetic-based guided positioning utilizes a magnetic attractor to guide a mobile device to a fixed location for accurate alignment. The advantage of this alignment method is its simplicity, but it has a limitation in that it can only charge in a fixed one-to-one position. Additionally, a magnetic material must be included in the charging device, leading to power loss due to eddy currents and an increase in temperature.

The free positioning method, which involves a movable primary coil, is shown in Figure 1.12(b). This method requires a primary coil that can be adjusted to align precisely with the charging device using a mechanically movable mechanism. The implementation of this method is straightforward when the charging pad is designed to charge only one device. However, charging multiple devices requires more complex motor control of the primary coil, making the system less reliable due to the presence of moving mechanical parts.

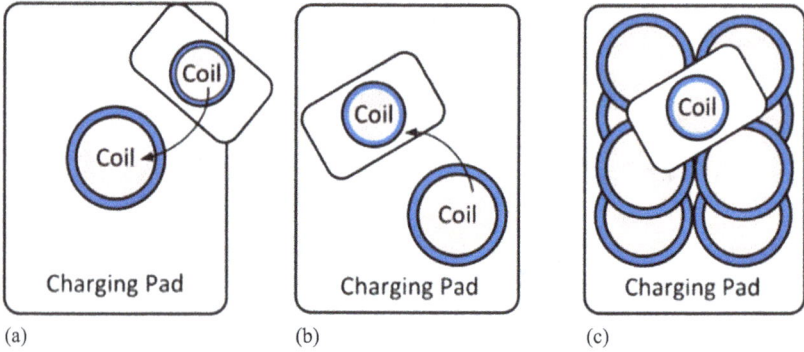

Figure 1.10   Models of alignment for wireless charging systems [4]. (a) Magnetic attraction, (b) moving coil, and (c) coil array.

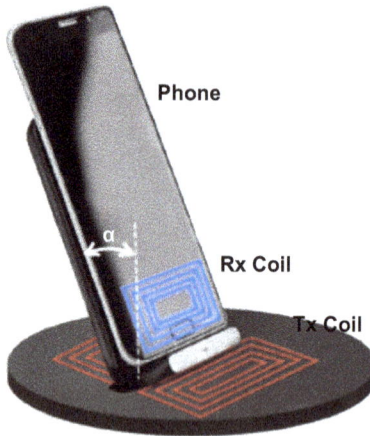

Figure 1.11   Vertical WPT system for portable device charging [5]

The use of free positioning based on an array of coils allows for the simultaneous charging of multiple devices in different locations, as demonstrated in Figure 1.12(c). This alignment approach offers greater user-friendliness than the other two approaches but requires a more complex winding structure and control electronic elements.

1.4.2   Vertical charging

'Vertical charging' as shown in Figure 1.11 is useful in many applications, including cell phone charging. In order to enhance the flexibility of wireless charging for objects placed vertically, the transmitter end employs an eight-shaped coil structure design [5].

Within the eight-shaped coil, current flows clockwise in one part of the feed loop and counter-clockwise in the other. These two segments of the eight-shaped coil generate opposing magnetic fields, resulting in a bipolar structure. Figure 1.12(a) shows the eight-shaped coil structure, where the red area represents the north pole and the blue area represents the south pole. Consequently, magnetic flux lines exit from one side of the eight-shaped coil Tx, pass through the vertically oriented receiving coil Rx and enter the opposite side of the Tx coil, creating a closed loop. Figure 1.12(b) shows the diagram depicting the distribution of the magnetic field in the vertical WPT system.

Figure 1.13 shows the mutual coupling and efficiency between the coils at different tilt angles of the receiver coil, where M_{13} represents the mutual

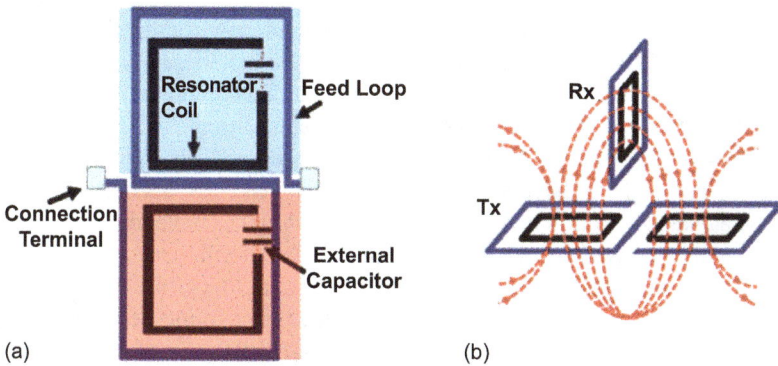

Figure 1.12   *Diagram of (a) the eight-shaped coil and (b) its magnetic field distribution [5]*

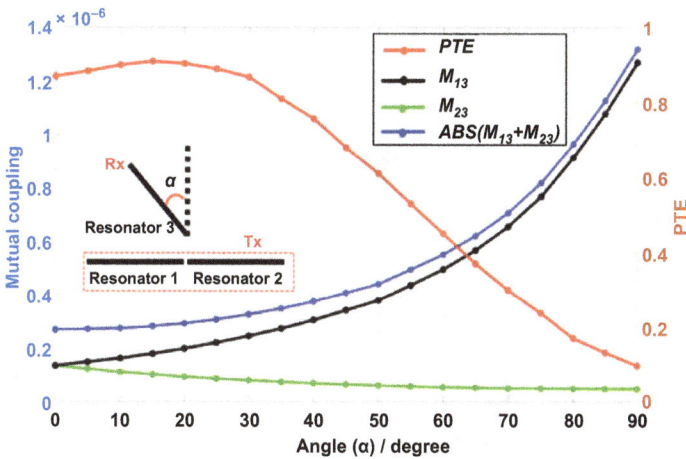

Figure 1.13   *Power transfer efficiency (PTE) and mutual coupling of an eight-shaped coil WPT system against angular misalignment α [5]*

inductance between one side of the eight-shaped coil Tx and the receiver coil Rx, and M_{23} represents the other side. It can be observed that when the receiver coil is tilted at an angle α of $\pm 40°$, the sum of the mutual inductance between the receiver coil and the eight-shaped coil ABS ($M_{13}+M_{23}$) does not vary significantly, resulting in an efficiency of over 70% within this range. Thus, the charging efficiency remains high even if the device is tilted at various angles, as long as the tilt angle does not exceed 40°.

1.4.3   Quadcopter charging

A quadcopter is an unmanned aerial vehicle (drone) that is powered and lifted by four rotors. These devices are utilized in numerous areas such as surveillance, search and rescue operations and object detection. However, due to the high power demand during flight, the battery life of a quadcopter is usually limited to only 20–30 minutes. WPT can be utilized to recharge the batteries. This method eliminates the need for a physical connection or manual intervention, although it might have a lower transfer efficiency compared to wired charging.

To ensure efficient WPT over a relatively large gap, it is imperative that the transmitter and receiver have a strong coupling. Nonetheless, accurately landing a drone on a charging station or device can be difficult. If the transmitter and receiver are misaligned, power loss occurs, reducing the overall efficiency of the charging system. Physical sensors such as ultrasonic and infrared types are utilized to align the drone with the charging station and maximize the coupling between the transmitter and receiver. This enhances the efficiency of WPT.

Figure 1.14 depicts a WPT system for a quadcopter, which is based on a transmitting coil array [6]. The system is composed of the following components:

1.   Wireless power transmitter module: This module consists of a power supply circuit and a four-way movable transmitter coil array.

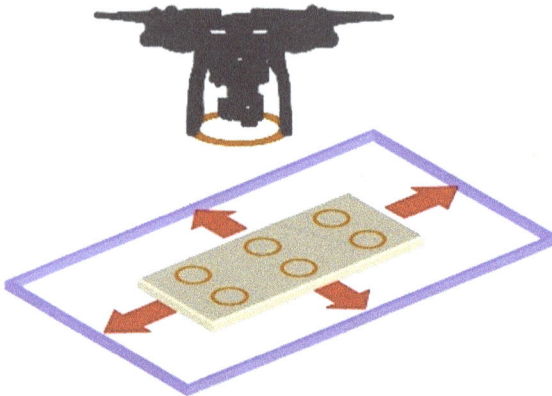

Figure 1.14   Wireless battery charging system for a quadcopter [6]

2. Wireless charging module: This is the device integrated into the unmanned aerial vehicle, and it includes a receiving coil and charging management circuitry.
3. Control unit: This module is responsible for determining the position of the drone on the charging pad through sensors and controlling the motors to move the coils to the optimal position.

1.5 WPT for EV charging

Wireless chargers enhance the versatility of charging options by eliminating the restriction of having to be physically connected to a cable when charging an EV. This enhances the convenience of the charging process, allowing for it to occur during brief stops or even while the EV is in motion. There are three modes of operation for wireless charging: static, quasi-dynamic and dynamic, as shown in Figure 1.15. The optimal mode for a particular situation can be determined based on factors such as the use of the EV, the available infrastructure and safety issues.

1.5.1 Static WPT

Static WPT takes place when EVs are in a parked position with the engine off for continuous charging over a long period. This is a common mode in public or private parking spaces. A typical static WPT system for EV charging is shown in Figure 1.16. Communication needs to be established between the vehicle and the

(a)   (b)   (c)

Figure 1.15   Three wireless charging operation modes. (a) Static, (b) quasi-dynamic, and (c) dynamic.

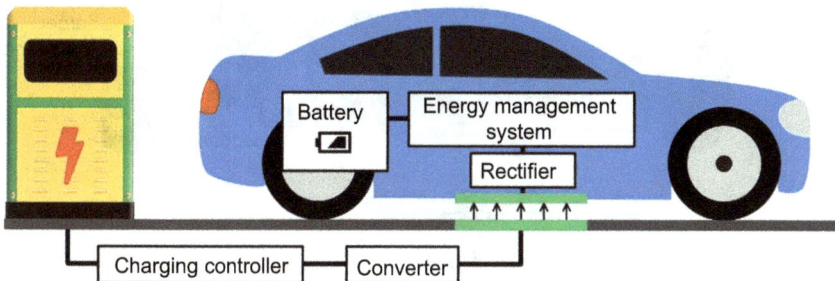

Figure 1.16   A typical static WPT system for an EV

charge controller to adjust charging parameters such as the start and end of charging, charging power, battery charge status, etc. Some complex alignment functions can be implemented into these chargers to minimize the effect of coil misalignment on charging efficiency. Usually, the maximum efficiency can be achieved only at the optimal position.

Static WPT offers a substitute for traditional plug-in charging methods for EVs, mitigating the potential safety hazards associated with connecting and disconnecting cables. This approach makes the charging process more convenient and hassle-free.

1.5.2   *Quasi-dynamic WPT*

Quasi-dynamic (stationary) wireless charging typically refers to the charging of a vehicle while it is in the stop-and-go mode. For instance, electrical energy is transferred to the battery of a public transportation vehicle through a pre-installed charging panel located under the ground when it stops at passenger stops or traffic lights. This type of charging helps replenish the battery during frequent stops, thereby increasing the range of the EV.

An example is the electric buses at Milton Keynes in the UK [7]. Electric buses are fully charged at night in designated parking areas. During daytime route operations, the electric bus will receive a booster charge at the beginning and end of its route services. As depicted in Figure 1.17, the bus stops over a charging plate installed beneath the road surface. The driver then lowers the receiving plate at the bottom of the bus to a distance of 4 cm from the road, and the bus is charged for a brief period before resuming its service.

Batteries take up a considerable amount of space and weight in an EV and are a major component of the cost. With quasi-dynamic wireless charging, the need for energy storage batteries in EVs can be reduced. The routes of public transport

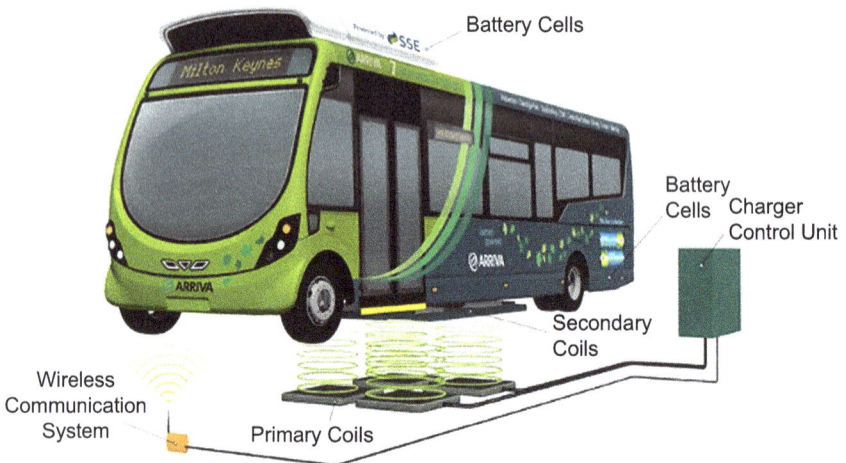

Figure 1.17   Wirelessly charged electric buses set for Milton Keynes [7]

vehicles are almost fixed and stop periodically at traffic signal points or pick-up and drop-off points for transport hubs, shopping centres, etc. The vehicle therefore only needs to carry a small capacity battery to move on to the next charging board before the battery is completely depleted. This reduces charging times and battery costs.

1.5.3 Dynamic WPT

Dynamic wireless power charging is considered the most challenging mode, as it enables EVs to be charged wirelessly while being driven on the road. This further reduces the size of the battery pack, providing greater convenience and flexibility for the vehicle. To implement dynamic charging systems, pre-programmed power transfer components such as coils should be buried under the road surface. Figure 1.18 shows two main designs: one features a segmented coil design, where each coil is powered by a separate power supply system, while the other uses a single coil design, which consists of a long track loop [8].

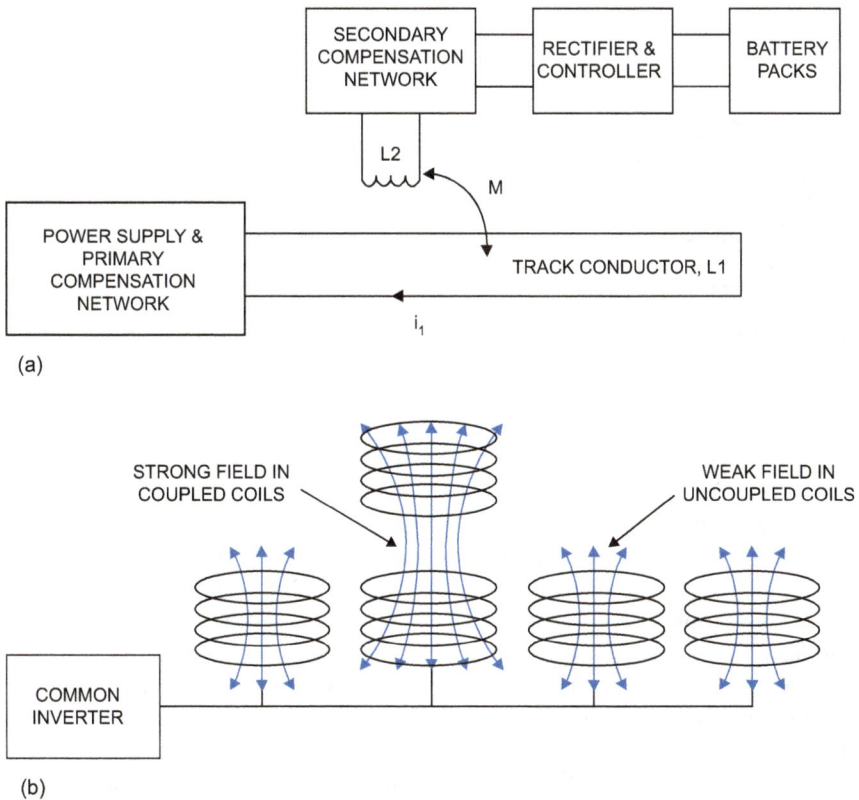

(a)

(b)

Figure 1.18   The dynamic charging systems with (a) single-coil and (b) segmented-coil structures [8]

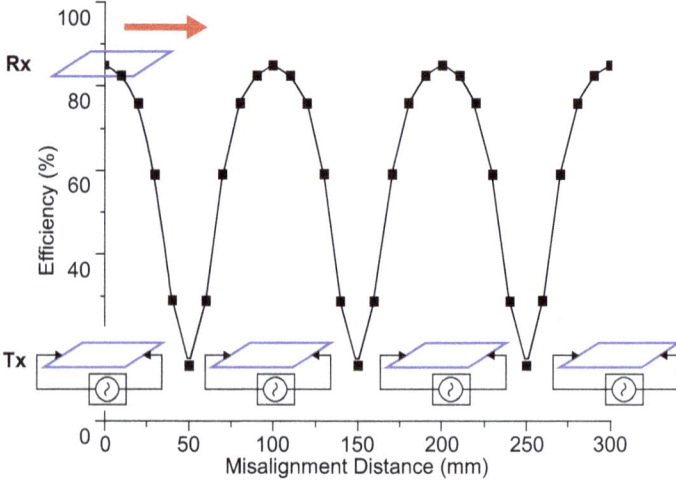

Figure 1.19   *Transfer efficiency when a receiver coil moves along a conventional segmented coil [9]*

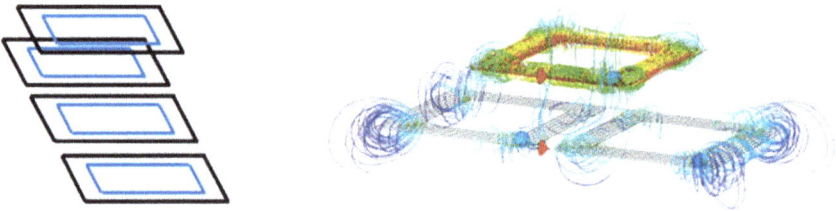

Figure 1.20   *Diagram of the conventional segmented-coil dynamic charging system [9]*

In the dynamic vehicle charging system based on a segmented coil design, maximum efficiency can only be achieved at the optimum position due to the constantly varying strength of the magnetic coupling between the transmitter and receiver. Figure 1.19 illustrates the variation in efficiency as the receiver coil moves along a transmitter coil array consisting of multiple independent coils [9].

A conventional segmented-coil dynamic charging system and its magnetic field distribution are shown in Figure 1.20. The receiving coil is placed in parallel with the transmitting coil. The transmitter coils are arranged along a straight line, with their magnetic field lines densely packed vertically in the centre of each coil. The magnetic field between the transmitter coils, however, is weak. As the receiver coils move along the path, the magnetic coupling cannot always be maintained at the optimum level, resulting in fluctuations in transfer efficiency.

A schematic of the dynamic charging system based on an eight-shaped coil and its magnetic field distribution is shown in Figure 1.21. Unlike the conventional

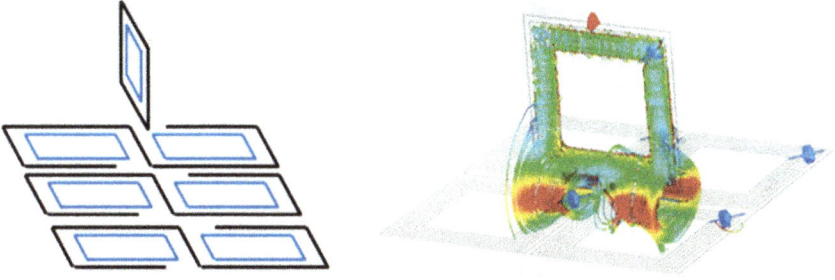

Figure 1.21   Diagram of the eight-shaped coil-based dynamic charging system [9]

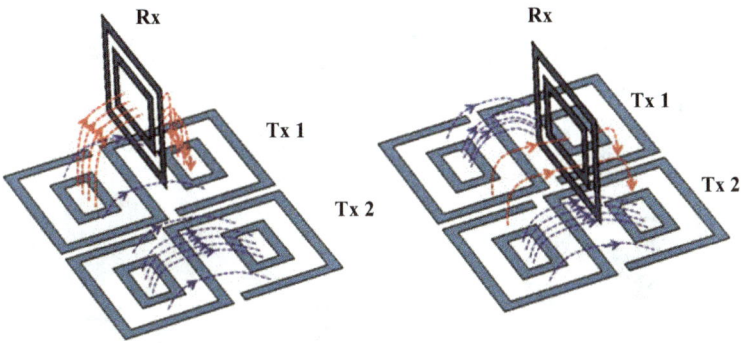

Figure 1.22   The magnetic coupling mechanism of an eight-shaped coil-based dynamic charging system [9]

method, the receiver and transmitter coils of this system are positioned orthogonally. The feed loop of the transmitter coil is based on the bipolar structure of the eight-shaped coil, allowing most of the magnetic flux to pass directly through the receiver coil for WPT. The magnetic field generated by the transmitter coil is distributed almost evenly along the travelling direction of the receiver, enabling the system to maintain a relatively constant efficiency. Moreover, placing the receiver vertically reduces eddy current losses on the metal parts of a vehicle due to the magnetic field distribution.

A simplified schematic diagram illustrating the distribution of the magnetic field during movement is presented in Figure 1.22. As the receiving coil moves, the coupling between the receiving coil Rx and the transmitting coil Tx_1 becomes weaker. Since the transmitting coil Tx_2 is already switched on by this time, the receiving coil is coupled to both the transmitting coils Tx_1 and Tx_2. Moreover, since the magnetic fields generated by the two transmitting coils are in the same direction, they can be superimposed, thereby enhancing the coupling between the coils.

The feed loops of multiple modules can be linked in series, as illustrated in Figure 1.23, to create a $1 \times n$ array of transmitting coils. The magnetic poles in the red region are the same, whereas those in the blue region are the opposite.

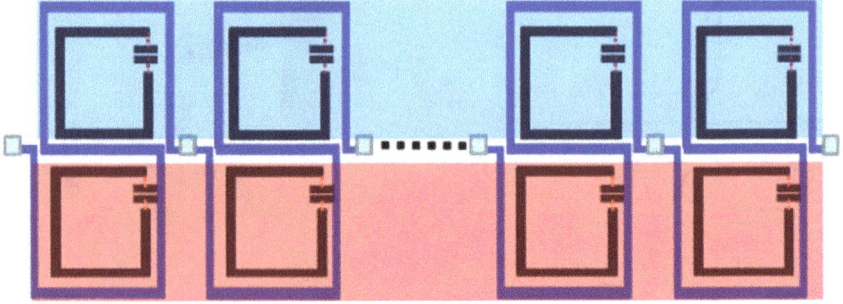

Figure 1.23   The multiple transmitting coil array [9]

Figure 1.24   Experimental setup of the eight-shaped coil-based dynamic charging system [9]

By increasing the number of transmitter coil modules, the charging area can be expanded further.

Figure 1.24 depicts the experimental setup of a dynamic charging system using a transmitting coil composed of three eight-shaped coil-based modules. The receiving coil is moved with 10 mm increments across a range of 0–210 mm. The transfer efficiency between the transmitting coil Tx and the receiving coil Rx is calculated from the measured S-parameters, as shown in Figure 1.25. When the Rx and Tx are aligned, the efficiency of the conventional structure is approximately 90%, while the efficiency of the eight-shaped coil-based system is around 70%. However, when the receiving coil is positioned between the two transmitting coils, the transfer efficiency drops significantly for the conventional system, while the efficiency of the eight-shaped coil-based system can be maintained at around 60%.

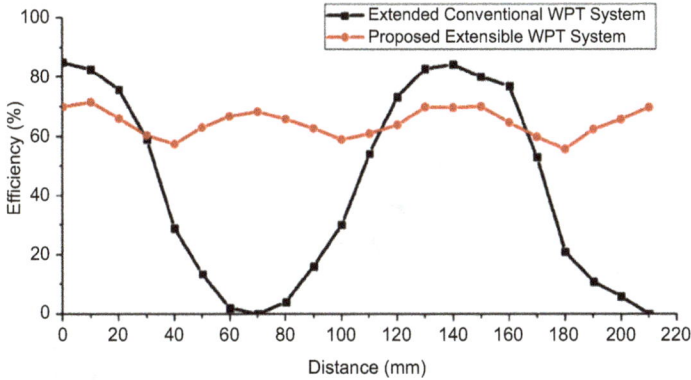

Figure 1.25   Comparison of the measured transfer efficiency of an eight-shaped transmitting coil and a conventional transmitting coil [9]

In other words, a conventional dynamic charging system can only achieve maximum efficiency when the Tx and Rx are perfectly aligned, whereas an eight-shaped coil-based system can maintain a constant efficiency of 60%–70% at all times.

1.6   WPT for implantable bioelectronics

Implantable electronic devices, based on bioelectronics, are used to regulate the function of organs by sending electrical signals. Such devices include pacemakers, cochlear implants, retinal implants and brain stimulators, as illustrated in Figure 1.26 [10]. When the energy stored in these devices is exhausted, they must be replaced through surgery, adding more challenges.

Three methods can be used to overcome battery life limitations. They comprise increasing battery capacity, powering the device from an external power source via an extended cord, and using wireless charging. However, the solution of using a larger battery capacity will increase the size of the battery. Thus, the size of the implanted device in the body will inevitably increase accordingly. The extension-cord-based solution requires routing the wires through the skin to the external power supply, posing a risk of infection and increasing the difficulty of follow-up care. Any accidental disconnection can be inconvenient or even life-threatening for the patient.

In contrast, a solution based on WPT and a small rechargeable battery can potentially solve these problems. WPT is becoming more common in implantable devices as a way to eliminate or extend battery life and reduce battery size. Rather than relying on batteries, WPT-powered devices only require an energy receiver. Energy receivers, such as coils, tend to be flat and thin, making implantable devices lighter and smaller. Some coils based on flexible circuitry can be more flexible and stretchable, allowing for improved flexibility and stretchability to better fit with human tissue.

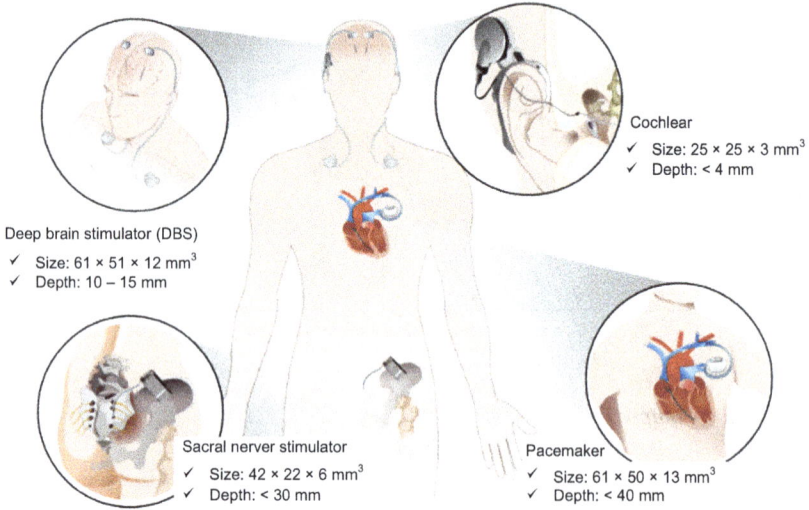

Figure 1.26   Graphical representation of various implantable medical devices and examples of sizes and insertion depths for each device: deep brain stimulator, cochlear, sacral nerve stimulator and pacemaker [10]

The use of WPT in implantable bioelectronics faces several unique challenges. The surrounding medium for implantable devices is human tissue, which is soft, moist, constantly changing and lacking in uniformity. The implant location and characteristics of the target organ can limit the design and size of implantable devices. Misalignment in position and orientation between the implanted device and the power transmitter can significantly reduce the effectiveness of WPT, resulting in weakened coupling between the transmitter and the device. Keeping the electromagnetic field strength within safe limits is also critical for implantable bioelectronics. One of the key indicators is the specific absorption rate, which measures the energy absorbed per unit mass of the body when exposed to electromagnetic fields. The standard states that the whole body average specific absorption rate should not exceed 0.4 W/kg and the local specific absorption rate should not exceed 10 W/kg [11]. It is necessary to adhere strictly to this standard to ensure the safe use of WPT in implantable bioelectronics.

1.6.1   Inductively coupled power transfer

The inductively coupled power transfer method is one of the most widely used WPT approaches for implantable bioelectronic devices. However, several challenges still remain. These include maximizing the power transfer efficiency to minimize the heat generated during the charging process and prevent damage to surrounding tissue, minimizing the size of the coil to reduce the overall size of the implanted device and ensuring that the magnetic field generated during charging does not interfere with the normal operation of other electronic devices or implanted devices in the vicinity.

The limited battery capacity of capsule endoscopes severely restricts their performance in vivo, including operating time, image resolution and the number of images they can capture. The use of WPT to power capsule endoscopes can be a promising solution to this problem. In the future, more complex and energy-consuming functions, such as directional control, are expected to be implemented. Figure 1.27 shows an inductively coupled near-field WPT system applied to a capsule endoscope with an array of wireless power transmitters installed beneath the floor of a test room, where patients wearing special jackets can move freely [12].

The system is based on two-hop inductive coupling. The first hop, using strong coupling to a high-Q resonator, allows for long-distance efficient power transfer from the power transmitter array to an open helix in the jacket. The second hop involves a loose coupling between the helix and the small-sized antenna in the endoscope capsule. The helix serves as a power relay, transferring power to the capsule. The antenna based on an LC resonator can be smaller than the conventional helix receiver at the same operating frequency, considering the capsule's strict size limitation. Through the combination of strong and loose couplings, the system can transmit power to a small receiver over middle-range transfer distances. The solution based on two-hop WPT can achieve efficiencies of up to approximately 30% over a transfer distance of approximately 1 metre, whereas conventional designs can only achieve similar efficiencies over a distance of a few centimetres. The wireless power transmitter array can be activated by a pressure sensor in its vicinity only when the patient is detected in an optimal position to improve the overall effectiveness of the system.

Cochlear implants enable acoustic perception by directly stimulating the auditory nerve in the cochlea with modulated electrical impulses, mimicking the

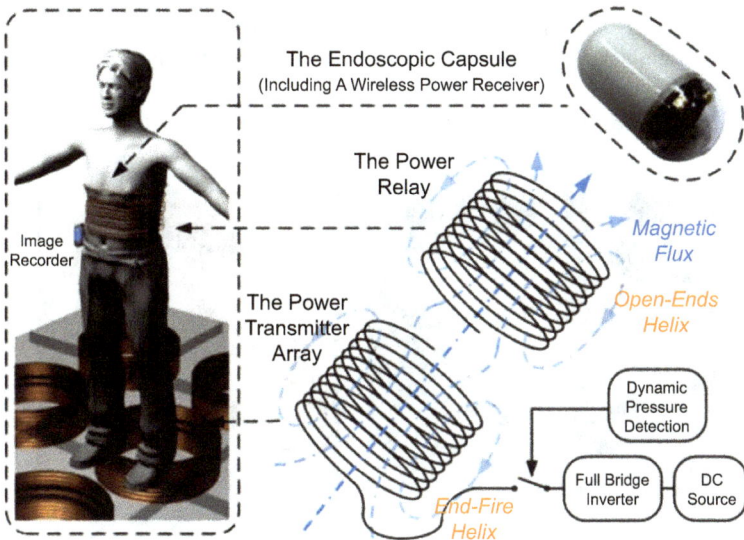

Figure 1.27 Wirelessly powered capsule endoscope system [12]

function of a healthy cochlea. By bypassing the damaged hair cells in the ear, the electrode array directly stimulates the inner ear sensory cells of the auditory nerve, restoring hearing in patients with conductive and sensorineural hearing loss. A wireless cochlear implant consists of an external device for processing sound signals and an implant unit, as shown in Figure 1.28 [13]. To ensure proper functionality of the cochlear implant, it is essential to maintain bi-directional communication between the external and implant units and ensure the stability of the WPT.

Figure 1.29(a) shows an implant for electrical stimulation of retinal neurons, which consists of an extraocular system and an implanted intraocular system [14]. Visual data is collected using the extraocular system and transmitted wirelessly via a coil-based antenna system. The intraocular system comprises an electrode array, a signal processing unit and secondary coils for power and data. The power coils with

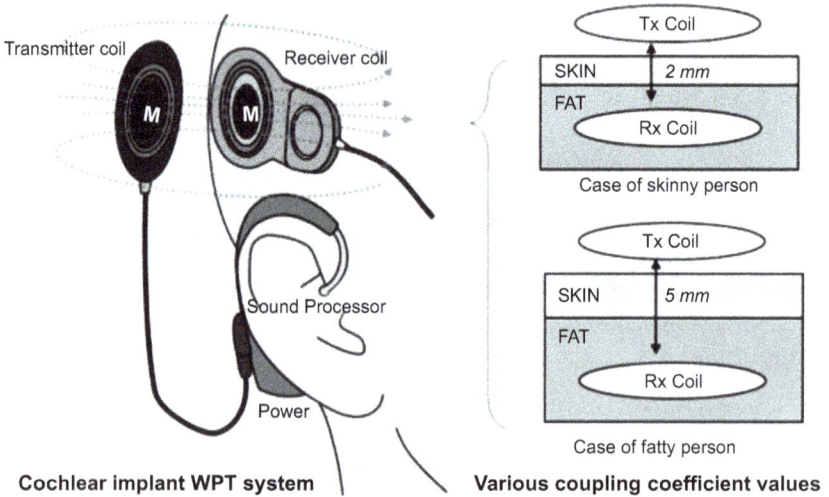

Figure 1.28   The cochlear implant with WPT [13]

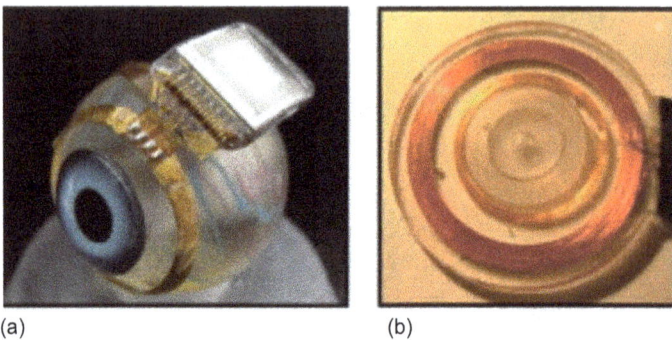

Figure 1.29   The electrical stimulation of retinal neurons (a) with WPT, (b) a sample of the primary coil [14]

a series resonant tank operate at a frequency of 125 khz in a typical design, and the RF power is converted to DC power within the implant by a dual half-wave rectifier. Stimulation data is transmitted using a carrier frequency of 15.5 MHz, which has a high data transfer rate and evades interference from power carriers of 125 khz and similar. The secondary coils consist of separate power and data windings made of 40 American Wire Gauge gold wire. To minimize the risk of tissue infection, the coils should be wound to match the curvature of the eye. In Figure 1.29(b), a sample of the primary coil is shown, where the power coil has a radius of 19 mm and the data coil has a radius of 12.5 mm.

1.6.2 Capacitively coupled power transfer

Capacitively coupled WPT systems using air as the medium have been intensively researched, but they can also be applied to other media such as skin, muscle and fat. This makes them suitable for WPT in implantable bioelectronics. The basic structure consists of four plates, two of which are placed inside the body to function as energy-receiving electrodes and the other two being located outside the body as energy-transmitting electrodes. The parallel plates, which are straightforwardly made of conductors, have a simple structure and are low-cost, making them ideal for implantable devices. Furthermore, capacitive-coupling-based WPT is less susceptible to misalignment and angular rotation. It offers greater design flexibility. Additionally, the system can still effectively transmit power even when integrated on a flexible substrate, which is crucial for implantable bioelectronics applications.

A scheme for wireless power supply for subdermal implants based on capacitive coupling transfer was proposed by Jegadessan *et al.* [15]. The system uses the skin as a dielectric material that separates two conductive surfaces, with power transmitted through the mutual capacitance between the conductors, as Figure 1.30 shows. An alternating electric potential is applied to the skin, generating an alternating electric field that penetrates the tissue and couples to the conductive plate under the skin, requiring two capacitors to complete the current loop. The subcutaneous implant's rectifier then converts the AC power to DC in order to power the device. Figure 1.31 shows the tissue model consisting of a skin layer, a fat layer and a muscle layer, with the transmitting patch on the outer side of the skin and the implanted receiving patch placed at the skin–fat interface. It makes the skin layer the main medium between the patches. The system can deliver 100 mW of power with 56% peak efficiency. Even in a bent state, up to 30% efficiency can be achieved. The capacitance between the patches is insensitive to bending, making capacitively coupled WPT suitable for biological applications. However, the limited distance between the transmitting and receiving patches, being around 2 mm, restricts the use of this system for powering implantable devices located in deeper organs.

Capacitive coupling systems typically use metal plates. However, these materials can be less biocompatible and tend to be stiff. In contrast, capacitive coupling systems based on ion-conductive plates offer improved biocompatibility and mechanical flexibility. Figure 1.32 shows an ion WPT system based on a

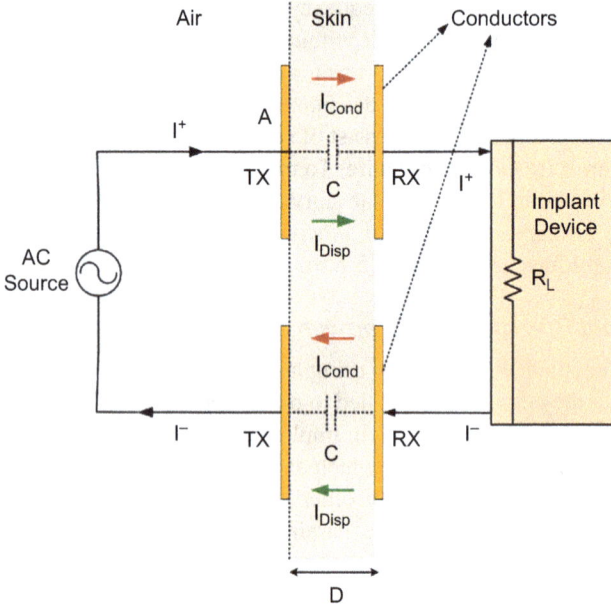

Figure 1.30   Schematic of the capacitively coupled WPT for subdermal implants [15]

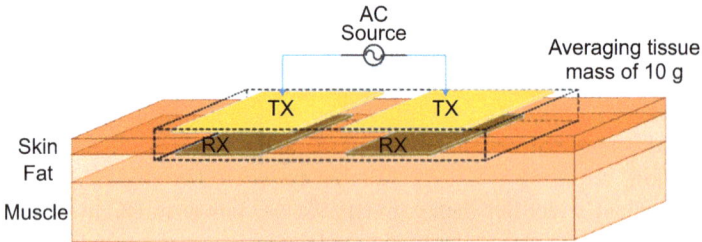

Figure 1.31   The cross-sectional view of the three-layer tissue model [15]

hydrogel receiver. The transmitter is made of metal, while the receiver is comprised of a hydrogel filled with sodium chloride solution [16]. Typically, the gaps between conductive plates are filled with a dielectric material to increase the capacitance. However, for implantable applications, the gaps are filled with tissues that contain electrolytes. To prevent direct contact with the gap-filling material and ensure insulation, the transmitter and receiver are covered with a Teflon film. An AC voltage source and a compensating inductor are connected to the metal plate-based transmitter. A rectifier and a resistive load are connected to the hydrogel receiver. The plates are positioned parallel to each other, forming a capacitor. When a voltage is applied between the transmitter and receiver, the charge in the transmitter

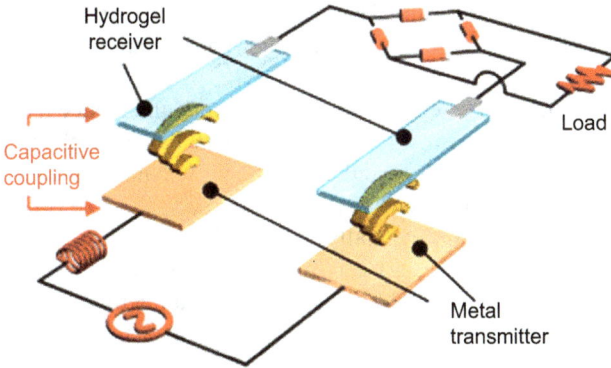

Figure 1.32   Basic principles and operations of an ionic WPT system [16]

attracts ions with an opposite charge in the receiver and repels ions with the same charge. The application of an AC voltage results in the alternating charge and discharge of the coupling capacitor, leading to a flow of current through the receiver. The compensating inductor, in combination with the parallel plate capacitor, induces an inductor–capacitor resonance, allowing for increased transfer of current through the capacitor.

1.7   Summary

This chapter focuses on near-field WPT technologies for transmitting power across small gaps. Two types of near-field WPT technologies are covered: inductively coupled WPT using magnetic fields and capacitively coupled WPT using electric fields. While the inductive coupling is likely to remain dominant in many mainstream applications in the future due to its maturity and efficiency. The capacitive coupling will continue to develop and could become dominant in areas where its unique advantages provide significant benefits. The future will likely see a coexistence of both technologies, each serving the needs of different applications and industries.

Additionally, far-field WPT technologies based on microwave radiation are briefly described. Near-field WPT has been widely used to charge small devices such as smartphones and electric toothbrushes. However, there are still some challenges, such as efficiency and distance limitations, particularly for vertical charging of mobile devices and quadcopter charging. The charging distance is usually comparable to the size of the transmitter. Several different design methods are discussed to address the coil alignment issue to optimize wireless charging performance. Near-field WPT is already available for charging EVs in static and quasi-dynamic modes, but a dynamic mode that allows vehicles to be charged while in motion remains challenging. For EVs and other similar applications, efficiency improvement would be the priority due to the high power involved. A few specific applications of near-field WPT in capsule endoscopes, cochlear implants and retinal implants are also described in this chapter. For these applications, the level of the received power is a major consideration point for health and safety reasons.

WPT technology can extend the life of the battery and reduce its size, resulting in smaller and lighter implants. A breakthrough in battery technology with significantly higher energy density would transform the landscape of WPT technology. It would reduce the frequency of charging, push the development of higher power transfer solutions and potentially shift the focus towards making WPT systems more efficient and capable of handling larger energy transfers. The potential of near-field WPT for implantable bioelectronics is attracting increasing interest. While far-field WPT is likely to become more important for specific applications like remote sensing, IoT devices and certain medical applications, near-field WPT will remain dominant in consumer electronics, high-power applications and scenarios where efficiency and alignment are critical. Powering aircraft using far-field WPT is unlikely with current technology due to the high power requirements and efficiency challenges. Advances in both WPT technologies and battery storage will shape the future landscape.

References

[1]   T. Imura, *Wireless Power Transfer: Using Magnetic and Electric Resonance Coupling Techniques*. Springer, Singapore, 2021.

[2]   C. Panchal, S. Stegen, and J. Lu, Review of the static and dynamic wireless electric vehicle charging system, *Engineering Science and Technology, An International Journal*, vol. 21, no. 5, pp. 922–937, 2018.

[3]   L. Dutta and F. Sumi, Future with wireless power transfer technology, *Journal of Electrical & Electronic Systems*, vol. 7, p. 279, 11/05 2018, doi:10.4172/2332-0796.1000279.

[4]   X. Lu, P. Wang, D. Niyato, D. I. Kim and Z. Han, Wireless charging technologies: Fundamentals, standards, and network applications, *IEEE Communications Surveys & Tutorials*, vol. 18, no. 2, pp. 1413–1452, 2016, doi:10.1109/COMST.2015.2499783.

[5]   C. Xu, Y. Zhuang, C. Song, Y. Huang and J. Zhou, Dynamic wireless power transfer system with an extensible charging area suitable for moving objects, *IEEE Transactions on Microwave Theory and Techniques*, vol. 69, no. 3, pp. 1896–1905, 2021, doi:10.1109/TMTT.2020.3048337.

[6]   A. Rohan, M. Rabah, M. Talha, and S.-H. Kim, Development of intelligent drone battery charging system based on wireless power transmission using Hill climbing algorithm. *Applied System Innovation*, vol. 1, p. 44, 2018, https://doi.org/10.3390/asi1040044.

[7]   BBC 2014. Wirelessly charged electric buses set for Milton Keynes. Retrieved 2023, from: https://www.bbc.co.uk/news/technology-25621426.

[8]   Z. Bi, T. Kan, C. C. Mi, Zhang, Z. Zhao, and G. A. Keoleian, A review of wireless power transfer for electric vehicles: prospects to enhance sustainable mobility. *Applied Energy*, vol. 179, pp. 413–425, 2016.

[9] C. Xu, Y. Zhuang, A. Chen, Y. Huang, and J. Zhou, Charging area extensible wireless power transfer system with an orthogonal structure. *2019 IEEE Wireless Power Transfer Conference (WPTC)*, London, UK, 2019, pp. 484–487, doi:10.1109/WPTC45513.2019.9055696.

[10] S. Ahn, H. Woo, K. Kim *et al.*, An out-of-phase wireless power transfer system for implantable medical devices to reduce human exposure to electromagnetic field and increase power transfer efficiency. *IEEE Transactions on Biomedical Circuits and Systems*, vol. 16, no. 6, pp. 1166–1180, 2022, doi:10.1109/TBCAS.2022.3222011.

[11] IEEE, in IEEE Standard for Safety Levels with Respect to Human Exposure to Radio Frequency Electromagnetic Fields, 3 KHz to 300 GHz, 2006, pp. 1–238.

[12] T. Sun, X. Xie, G. Li, Y. Gu, Y. Deng and Z. Wang, A two-hop wireless power transfer system with an efficiency-enhanced power receiver for motion-free capsule endoscopy inspection. *IEEE Transactions on Biomedical Engineering*, vol. 59, no. 11, pp. 3247–3254, 2012, doi:10.1109/TBME.2012.2206809.

[13] S. Hong, S. Jeong, S. Lee, *et al.*, Cochlear implant wireless power transfer system design for high efficiency and link gain stability using a proposed stagger tuning method. *2020 IEEE Wireless Power Transfer Conference (WPTC)*, Seoul, Korea (South), 2020, pp. 26–29, doi:10.1109/WPTC48563.2020.9295642.

[14] S. K. Kelly, D. B. Shire, J. Chen *et al.*, A hermetic wireless subretinal neurostimulator for vision prostheses. *IEEE Transactions on Biomedical Engineering*, vol. 58, no. 11, pp. 3197–3205, 2011, doi:10.1109/TBME.2011.2165713.

[15] R. Jegadeesan, K. Agarwal, Y.-X. Guo, S.-C. Yen and N. V. Thakor, Wireless power delivery to flexible subcutaneous implants using capacitive coupling. *IEEE Transactions on Microwave Theory and Techniques*, vol. 65, no. 1, pp. 280–292, 2017, doi:10.1109/TMTT.2016.2615623.

[16] C. C. Kim, Y. Kim, S. H. Jeong, *et al.* An implantable ionic wireless power transfer system facilitating electrosynthesis. *ACS Nano*, vol. 14, no. 9, pp. 11743–11752, 2020, doi:10.1021/acsnano.0c04464.

Chapter 2

Magnetic confinement fusion power

Howard Wilson[1]

2.1 Fusion energy basics

2.1.1 Fusion power and its benefits

Fusion is the process that powers the sun. It occurs when the nuclei of light atoms are brought sufficiently close together that they fuse to create a heavier nucleus and release a large amount of energy. It is this energy that we seek to harness to generate electricity to power our homes, industry and transport.

The easiest fusion reaction to achieve is between two isotopes of hydrogen – deuterium (D) and tritium (T). When they fuse, they create a helium nucleus and a neutron. The fusion energy that is released in the reaction is contained within the kinetic energy of the products: 80% of it in the neutron and 20% in the helium nucleus. Such an energetic helium nucleus is also called an alpha (α-) particle. It is confined with the DT, giving up its energy to help keep the fuel sufficiently hot to maintain the fusion reactions. The neutron escapes the fuel, and its energy is captured in a blanket structure to create electricity (for example). While most fusion energy schemes target this DT reaction, some consider alternatives that do not produce neutrons. These energetic neutrons can challenge materials integrity and have implications for waste management, so there are advantages to aneutronic schemes. One alternative, for example, is the reaction between a proton and a boron nucleus to produce three alpha particles. However, most believe that the first fusion power plants will operate with DT, so we restrict consideration to this reaction here.

Deuterium is plentiful in nature, but tritium decays with a half-life of just over 12 years and is therefore very rare. The strategy to address this is to manufacture – or 'breed' – the tritium on-site as it is required. An initial charge of tritium is used to start up the power plant. As conditions approach those for fusion to occur, energetic neutrons are produced at 14 MeV. These must be captured in the above-mentioned blanket structure, which serves two purposes: (1) to extract the energy from the neutron for conversion into electrical power (e.g. via standard turbine technology) and (2) to react the neutron with a lithium nucleus, which then produces tritium. To be specific, the reaction between a neutron and a lithium-ion

[1]Fusion Energy Division, Oak Ridge National Laboratory, USA

produces a tritium ion and helium. This tritium should then be extracted from the blanket to be used in the fusion process.

An important concept is the 'tritium breeding ratio', or TBR, which measures how many tritium ions are manufactured, on average, for each neutron produced in the fusion process. To be self-sustaining brings a necessary condition of TBR>1. This is challenging because not every neutron produced in the fusion reactions will react with lithium – some will interact with the plant structure, for example. Thus, the blankets are complex components, requiring an appropriate mix of materials to multiply neutrons and moderate their energy, as well as (potentially) an optimised distribution of the two naturally occurring isotopes of lithium, which have different interaction cross sections with different energy neutrons.

Note that our discussion so far demonstrates that while our fusion reaction is between deuterium and tritium, the raw materials for the fuel are deuterium and lithium, both of which are plentiful – sufficient to power civilisation for thousands of years. There are other major benefits to fusion as a sustainable energy source:

- No greenhouse gases are produced in the process.
- Fusion is inherently safe. The conditions required for fusion are difficult to achieve, so if anything goes wrong, the process simply stops – it is not possible to get a runaway reaction from fusion.
- Fusion power will be delivered on demand and close to where it is required, with no dependence on weather conditions and regional climate.
- The high power density of fusion energy means it takes up minimal land, and can be used to power large mega-cities around the world.
- Radioactive waste is produced, but its level is likely to be categorised as 'low' and can readily be managed in a safe and environmentally sustainable way.

It is worthwhile commenting a little more on the waste, which is a rather subtle and complex topic. When a neutron interacts with certain materials, it can create radioactive elements. Thus, the amount of activated waste from a fusion plant and its level depends critically on the materials used for the construction. Novel materials are being explored and developed, but many of the structural components will likely be made from steels, which are dominated by iron. The interaction between iron and fusion neutrons creates very short-lived isotopes, so the associated waste is negligible. However, some of the trace elements that are added to the iron to give any particular steel its mechanical properties can result in significant waste. Nevertheless, the small fraction of such elements, and the ability to tailor their content to produce 'low activation' steels, mean that this waste from future commercial fusion plants will likely be categorised as 'low level' (not exceeding 4 GBq per tonne of alpha activity, or 12 GBq per tonne of beta/gamma activity) and relatively short-lived (up to 100 years). Importantly, this level of waste does not require advanced technical solutions for its storage. The take-home message is that the waste produced by a fusion power plant is readily managed and furthermore can be minimised by careful choice of the materials used and through de-tritiation technologies.

2.1.2   Fusion energy – the conditions

We understand the conditions required to produce fusion reactions very well – the research question that the international fusion community has been addressing since the 1950s is 'How do we achieve those conditions in a controlled manner here on Earth?'. In this section, we will explore them in more detail, introducing a key quantity called the fusion triple product.

The first parameter we introduce is the density – the number of deuterium and tritium ions per unit volume. Clearly, the rate of fusion reactions will be proportional to the density of both the deuterium and tritium: the more fuel nuclei that are packed into a given volume, the more chance they have of coming close enough to fuse and the higher the fusion power released. The optimal situation is when there are equal densities of deuterium and tritium, although the maximum is quite broad, and even a 25:75 mix of D:T produces a fusion power of only 25% less than the optimal 50:50 mix.

The second parameter is the temperature of the D-T fuel ions. To explore this, we introduce the concept of 'reactivity', $\langle \sigma v \rangle$, which is related to the probability of two ion species having a fusion reaction. The fusion power per unit volume is then given by

$$\bar{P}_{fus} = 5n_D n_T \langle \sigma v \rangle E_\alpha \qquad (2.1)$$

where n_i is the density of species i, and E_α = 3.5MeV is the energy of the alpha particle produced in the fusion reaction.

The reactivity depends on both the elements involved in the reaction and their temperature. Figure 2.1 shows this dependence for three different fusion reactions, measuring temperature of the ions in units of keV, where 1 keV is equivalent to 12 million Kelvin. Important features of Figure 2.1 include the following: (1) The DT reactivity is at least a factor of ten larger than for other fusion reactions shown: deuterium only (DD) and deuterium with the helium-3 isotope; (2) The optimal temperature for the DT reaction is in the region of 200 million Kelvin. This

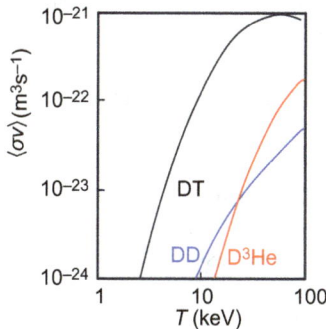

Figure 2.1   Reactivity, $\langle \sigma v \rangle$, for three different fusion reactions as a function of ion temperature, T

challenging temperature is approximately ten times that at the centre of the sun, and provides a first quantification of the conditions required for fusion power.

The third parameter to define the conditions for efficient fusion is the confinement time. This is a quantity that measures the effectiveness of the confinement system – the faster that thermal energy and particles leak from the confinement system, the shorter the confinement time. Mathematically, one measure of confinement time (the energy confinement time, τ_E) is defined in terms of the rate of change of thermal energy of the DT fuel, W:

$$\frac{dW}{dt} = P_{tot} - \frac{W}{\tau_E} \qquad (2.2)$$

where t is the time and P_{tot} is the total heating power provided: externally applied by the user and the heating provided by the fusion α-particles. From (2.2), it can be seen that if we switch off the heating power, then the plasma cools down exponentially at a rate inversely proportional to τ_E, demonstrating that τ_E represents how effective the confinement system is. In steady state, (2.2) provides a convenient definition of the confinement time:

$$\tau_E = \frac{W}{P_{tot}} \qquad (2.3)$$

We can give a physical picture of the confinement time by considering an analogy with a bucket of water. For a perfect bucket (our water confinement system), the water level will not change – it has an infinite 'water confinement time'. If we now drill a number of holes into the bucket, water will leak out. The water confinement time is the time taken for a fraction $1/e$ of the water to leak out (where $e = 2.718$). The more holes we drill, the faster the water will leak and the shorter the confinement time. We see that the confinement time is directly related to how leaky the bucket is. However, we can maintain the level of water if we pour some into the bucket at the same rate that it leaks through the holes. Thus, two things are clear from our model system: the shorter the confinement time (1) the more water has to be poured in to maintain the level, and (2) the more water escapes through the holes. Returning now to our fusion system, the level of water is analogous to the thermal energy in the fuel, and the amount of water added is analogous to the heating power. Thus, the shorter the energy confinement time the more heating power must be added, and the more thermal energy is exhausted. We will see later that managing the exhaust power is a major challenge for magnetic confinement fusion, and this has a direct link to the effectiveness of the confinement system. Confinement is a crucial aspect of fusion research, which we shall return to later.

Recall that the total power provided to the fuel is the sum of any external heating power, P_{ext}, and the power associated with the fusion alpha particles, $P_\alpha = P_{fus}/5$ (i.e. 3.5MeV of the total of 17.5MeV produced by each D-T fusion). A case of particular interest is when the fusion process is sustained by the alpha particle heating alone, with no need for additional heating power. We call this situation 'ignition'. Under certain assumptions about the distribution of fuel density and temperature, one finds that the condition for achieving ignition is provided by the so-called Lawson condition. This can be derived by assuming a steady state, and combining (2.1) and (2.3), to provide

$$n_i \tau_E T_i > 5 \times 10^{21} \text{m}^{-3} \text{keVs} \qquad (2.4)$$

The product of ion density, n_i, temperature, T_i and energy confinement time, τ_E, is called the triple product. Measuring the triple product of the fusion fuel and comparing it with the value required for the Lawson criterion from (2.4) provides a key quantitative measure of how close the conditions are to those required for fusion. We will return to this later.

2.1.3 Plasma

Three states of matter – solid, liquid and gas – are very common and familiar to us here on Earth. Elements transition from one state to the next by the introduction of energy, typically in the form of heat. As an example, consider heating ice, which then melts to become water and then boils to become steam. If one adds progressively more heat, the water molecules would dissociate into atoms and then the electrons would be released from the ions to create the state of matter called plasma. A plasma consists of many charged particles – for example, positively charged ions and negatively charged electrons. The positive and negative charges balance, so overall the system is charge-neutral. Nevertheless, the fact that the particles within a plasma are charged means they interact with each other, and with electromagnetic fields, in very different ways than the interaction between the particles of a neutral gas (e.g., the air in a room). A consequence of this is that plasmas behave differently to neutral gases, such as the rich variety of waves they can support. For this reason, plasmas are categorised as the fourth state of matter.

Natural plasmas are rare on Earth, but examples include lightning. However, in the universe, it is the most abundant state. The sun, and indeed all stars, are in the plasma state, as is space and the interstellar medium. The aurora borealis ('northern lights') is a most beautiful example of a plasma-based phenomenon as a visible display of the waves that can occur when plasmas interact with a magnetic field – in this case, the Earth's magnetic field in a phenomenon stimulated by energetic particles streaming from the sun. On Earth, plasmas are mostly made by humans for various technological applications. In manufacturing, 'low-temperature plasmas' are used for advanced coating technologies and etching fine structures in semiconductor chips, while so-called atmospheric plasmas show great promise for biomedical applications. The high temperatures required for fusion mean the DT fuel is also in this plasma state. Understanding how to achieve the conditions for fusion to occur, as provided by (2.4), therefore requires an understanding of plasma physics. Specifically, understanding the confinement time achievable in a particular configuration requires a detailed understanding of the interaction between the many waves that can be supported by plasmas, and the deuterium and tritium ions. These can drive fine-scale instabilities and turbulence that degrade the confinement time – a complex and rich area of physics that is fundamental to the achievement of controlled fusion.

2.2   Confinement systems

Broadly, there are two approaches to achieving the required confinement time – magnetic and inertial confinement. In inertial confinement, a small spherical pellet of frozen DT fuel, a few mm in diameter, is encapsulated in a thin coating of a heavier material. High-power lasers (for example) are simultaneously focused on the pellet, either directly or via a small cavity called a Hohlraum which, when heated by the high power lasers, bathes the pellet in X-rays. The coating rapidly ablates, pushing against the DT fuel to compress it to extremely high density, some one thousand times higher than standard solid density and heating towards fusion temperatures as it is compressed. This compression is equivalent to taking a standard building brick and squeezing it down to the size of a Lego brick, albeit at a much smaller scale. The confinement time is very short – billionths of a second – essentially a consequence of the finite inertia of the fuel. The huge density is what enables the triple product to be approached despite this low confinement time. The process is illustrated in Figure 2.2.

Much of the scientific challenge of inertial fusion is related to maintaining an even compression of the pellet, which must be manufactured to extremely high precision, carefully tailoring the distribution of DT inside it. Technical challenges for commercial fusion delivery also include the development of efficient, powerful lasers able to deliver several pulses per minute and bringing the cost of manufacturing the fuel pellets down by five to six orders of magnitude compared to present-day costs. Novel approaches to inertial fusion, such as the projectile approach being developed by the private company First Light Fusion, aim to

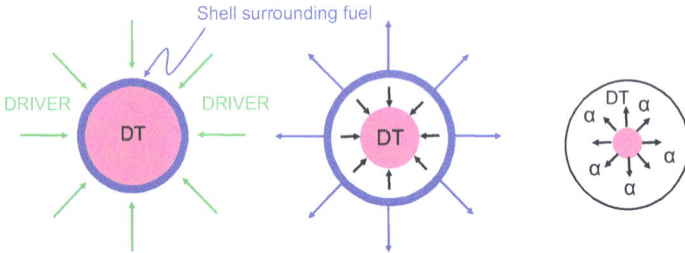

Figure 2.2   Cartoon illustrating the essentials of a particular inertial confinement fusion approach. Left: Frozen DT fuel is captured in a spherical shell, which is rapidly heated by a 'driver' (e.g. direct drive using high-power lasers in this sketch). Middle: the shell ablates, pushing against the DT fuel to compress it to 1000 times solid density, heating it to fusion temperatures at the same time. Right: a central hot spot triggers fusion reactions, producing energetic alpha-particles; these heat up the outer parts of the DT fuel to fusion conditions, creating a so-called 'burn wave' through the plasma.

simplify the compression process and circumvent these technical challenges, but a clear pathway to commercial realisation is not yet defined.

The National Ignition Facility in the US is the world's most advanced inertial confinement fusion facility. It passed the 'scientific breakeven' condition in 2022, producing 3.15 MJ of fusion energy from just 2.05 MJ of injected laser energy [1]. While much more energy was required to power the lasers (so the result is some way short of the requirements for a commercial fusion power plant), this nevertheless provides exciting new plasma conditions with significant fuel heating provided by the fusion-produced alpha particles. It is a very important milestone on the pathway to fusion power, providing a solid benchmark of our understanding of the requirements.

Magnetic confinement fusion (MCF) employs a configuration of magnetic field to hold the DT plasma at the optimal fusion temperature, often in a toroidal geometry. It is somewhat more developed towards a commercial power source than inertial fusion energy. Indeed, viable MCF power plant designs based on toroidal plasma configurations are now in sight, although significant scientific and technological challenges remain. There are two components to the confining magnetic field. The dominant one is called the toroidal (component of) magnetic field, which is created by an arrangement of current-carrying coils, such as that shown in Figure 2.3. If the field is high enough, typically a few Tesla, then, to leading order, the ions at fusion-relevant energies spiral rapidly around the magnetic field lines in a tight helical path with a 'Larmor radius' of about a millimetre or two. Thus, one might expect that in the magnetic geometry shown in Figure 2.3, the ions will simply circulate around the toroidal direction and be extremely well confined away from the toroidal chamber walls. Unfortunately, that is not the case – the fact that the magnetic field lines are curved and that the magnitude of the magnetic field becomes smaller further from the symmetry axis causes the ions to drift vertically away from the toroidal surfaces containing the magnetic field lines. Electrons also follow a helical path along magnetic field

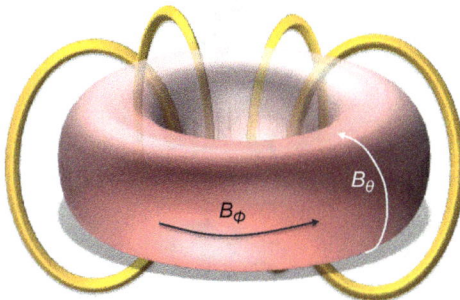

Figure 2.3   Sketch to illustrate how a toroidal component of magnetic field, B_ϕ, is generated from the yellow current-carrying coils. The poloidal component of magnetic field, B_θ, is also shown, which is required for effective confinement.

lines with an even tighter Larmor radius because of their lower mass compared to the ions. They also experience the same vertical drift, but because they have the opposite sign charge to the ions, they drift in the opposite direction. A vertical electric field is created as a consequence of the opposite vertical drifts of ions and electrons, which then causes an additional, so-called 'E cross B drift' of both particle species, away from the symmetry axis until they strike the vessel wall. The consequence is a short confinement time. While it is much larger than the confinement time of inertial fusion, the fuel density is many orders of magnitude less and so the triple product for this system is insufficient.

The effect of the vertical particle drifts can be eliminated by modifying the geometry of the magnetic field. Specifically, the field lines are formed such that they spiral over a toroidal surface. To achieve this requires an additional 'poloidal' component of the magnetic field, as well as the toroidal one (Figure 2.3). An example is the tokamak, the magnetic coil geometry for which is shown in Figure 2.4. In this figure, the blue D-shaped coils carry a current to create the toroidal component of the magnetic field. The green inner poloidal field coils create a solenoid; when the current is ramped in these solenoid coils, an electric field is induced into the plasma, which then drives a current in it by transformer action – a direct consequence of Faraday's law. A plasma is a highly conducting medium because of its free electrons and ions, which are rapidly accelerated by the electric field. This results in a so-called plasma current, which is in the toroidal direction, and is typically around a mega-Amp. This is sufficient to create the poloidal component of the magnetic field at significant fractions of a Tesla, but typically rather smaller than the toroidal component (with notable exceptions in the so-called 'spherical tokamak'). Another coil set which is illustrated in Figure 2.4 is the poloidal field coils. These act to control the plasma position and to shape its cross-section – a plasma with a D-shaped cross section has significantly better stability and confinement properties than the standard circular cross-section of a simple ring-doughnut shape.

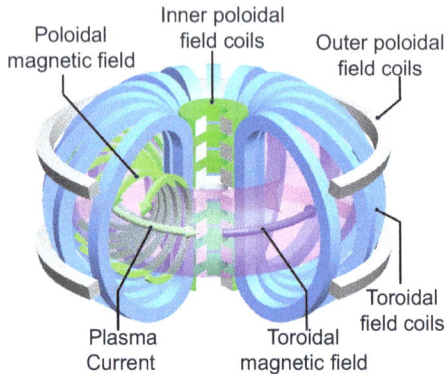

Figure 2.4   The magnetic coil set of a tokamak (adapted from a configuration by EUROfusion)

The net result of the full set of tokamak coils is magnetic field lines that spiral around (imaginary) toroidal surfaces called flux surfaces. An important parameter that characterises the magnetic geometry is the safety factor, q. This is defined as the number of revolutions a magnetic field line makes in the toroidal direction to perform one poloidal revolution. The flux surfaces form a set of toroidally symmetric nested tori, centred on the so-called magnetic axis, as shown in the poloidal cross-section of Figure 2.5. In standard operational scenarios, the safety factor increases from a number typically around $q=1$ at the magnetic axis to a value of 3 or more near the plasma edge (i.e. close to the surface labelled 'Separatrix'). Indeed, if q at the edge is too low, the plasma is susceptible to a particular kind of instability called the kink mode. This can lead to loss of plasma control and a sudden termination of the discharge in an event called a disruption. We shall return to the issue of disruptions later. Given that q at the plasma edge is proportional to the ratio of toroidal field to plasma current, kink mode stability then imposes a maximum current that can be driven in the plasma for a given toroidal field.

One challenge associated with a tokamak is the (apparent) need to ramp the current in the solenoid in order to sustain the plasma discharge. This means that in its most basic form, it is an inherently pulsed device, with pulse lengths lasting from of order a second up to many minutes, depending on the tokamak size and the technological infrastructure. A typical confinement time is much less than this, ranging from hundreds milliseconds on the smaller tokamaks up to a second or so

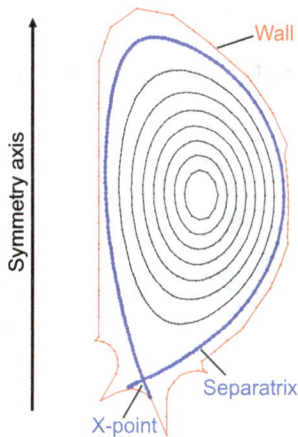

Figure 2.5   Sketch of poloidal flux surfaces in a tokamak plasma, shown in the poloidal cross-section. Magnetic field lines lie in these surfaces, spiraling over them poloidally and toroidally. The full 3D structure of the magnetic field is obtained by rotating the above cross-section about the vertical axis of symmetry shown. The hot plasma would be contained within the surface labelled 'Separatrix' and the red line provides a representation of where the material vessel wall could be positioned.

on the next generation, large tokamaks. This confinement time is billions of times larger than that achieved in inertial confinement fusion facilities, but that is off-set by the very much lower density of the plasma fuel.

There is another attractive option for magnetic confinement fusion, called the stellarator. The principle of a stellarator is fundamentally the same as a tokamak, in that the plasma is confined in a toroidal geometry by a combination of poloidal and toroidal magnetic field components. A key difference is that the poloidal component of the magnetic field is created from a coil set, adjusting the shape of the toroidal field coils so that both components can be generated (Figure 2.6). This removes the requirement for the plasma current, which then means there is no need for a solenoid, and so the stellarator can operate in a steady state. Moreover, because there is no need to drive current in a stellarator, this means it is not vulnerable to disruptions that are a concern for tokamaks. There are therefore significant benefits associated with a stellarator, but there are also challenges. One is that the magnetic geometry is more complex, as the stellarator does not have the toroidal symmetry of the tokamak. This means that generally energetic particles are not well confined in a stellarator, which then has an impact on their effectiveness for heating. However, careful design of the magnetic field coils to provide a specific magnetic geometry can recover the fast particle confinement. As a result, stellarators remain an attractive option for commercial fusion power, and are an active avenue of research.

To summarise, in this section, we have discussed some of the most promising approaches to delivering fusion power. Amongst these, the tokamak is presently the most advanced, so we will focus on this design for the remainder of this chapter to provide a specific reference design, while noting the importance of continuing to develop the full range of design options towards commercial fusion power.

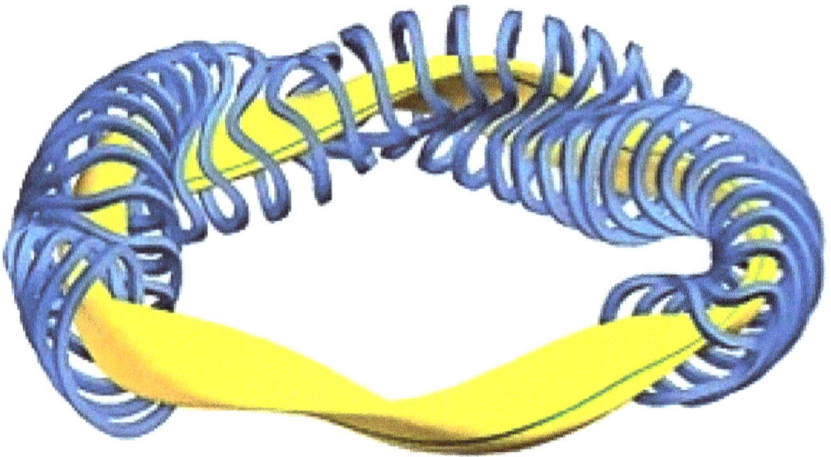

Figure 2.6   An example of the stellarator coil set (blue coils surrounding the plasma depicted in yellow), in this case associated with the Wendelstein 7-X device in Germany (figure taken from Wikipedia)

2.3 Tokamak physics

The tokamak is a Russian invention, which was first explored in the 1960s in a small experimental device called T3. Its early promise led to numerous other larger and more advanced tokamaks being built around the world. Figure 2.7 captures the huge progress that has been made in tokamaks, marching ever closer towards the plasma parameters required for commercial fusion power. It shows the values of triple product and temperature achieved by tokamaks since the first measurements were made on the T3 tokamak in the 1960s. Temperature is measured in units of keV, and the sun's core temperature of about 10 million Kelvin is shown for reference. The three green curves provide a measure of fusion power from DT reactions via the quantity Q. This is defined as the ratio of the fusion power produced to the heating power injected into the plasma to achieve the fusion conditions, and clearly, $Q \gg 1$ must be achieved for commercial relevance. Scientific breakeven corresponds to $Q=1$, but this is not sufficient for a commercially viable plant as one requires sufficient additional fusion reactions to overcome the power-hungry systems required for tokamak operation. An infinite value of Q means that the heating provided by the fusion-produced alpha-particles is sufficient to maintain the plasma in the conditions required for fusion, without any external heating. One can see that the green curves indicate an optimal temperature where the

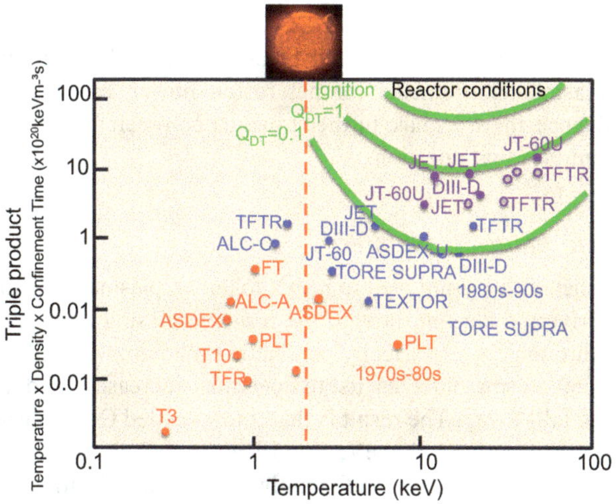

Figure 2.7   *Progress in tokamak physics towards the triple product and temperature required to realise commercial fusion power. The green curves identify the conditions to achieve certain values of Q for DT fusion; the names are different tokamaks, and the colours represent the conditions achieved in different decades: 1970s in red, 1980s in blue and 1990s in purple. (Adapted from a figure originally produced by EUROfusion, but no longer on their website.)*

required triple product is lowest, corresponding to about 20 keV, or two hundred million kelvin – ten times the temperature at the sun's core.

Figure 2.7 shows that the original results from T3, while ground-breaking at the time, were a factor of more than ten thousand below what is required in triple product, and a factor of a hundred below the optimal temperature. The T3 results were confirmed by the international community, which then drove excitement around the world and enthusiasm to build bigger and better tokamaks. This led to factors of ten improvement in both temperature and triple product during the 1970s. The 1980s saw a further factor of ten improvement in triple product, with temperatures now nudging up against the optimal for DT fusion. Finally, during the 1990s tokamaks, were not only achieving the required temperature for fusion, but were also approaching the triple product required for scientific breakeven in DT plasmas. However, most tokamaks are not designed to operate with tritium because that would result in higher construction and operational costs, as well as a more challenging environment in which to perform experiments. Rather, they typically operate in pure deuterium, which is sufficient to learn how to get close to fusion conditions. Two exceptions in Figure 2.7 are the US tokamak TFTR and the EU tokamak JET, both of which were designed to operate with tritium. TFTR is no longer operating, while DT experiments were most recently performed on JET in the early 2020s. These JET experiments produced a world record in fusion energy – 69MJ over about 5s [2] – which was in line with theoretical predictions. JET ceased operations at the end of 2023, leaving no currently operating DT-tokamak in the world.

While tokamaks have demonstrated great advances, they still fall somewhat short of the plasma requirements for net fusion power gain. So what more is required to advance the tokamak further towards commercialisation? The rest of this section examines that question.

2.3.1   Plasma heating schemes

A number of schemes are employed to heat a tokamak plasma to fusion conditions. For example, driving a current through the resistive plasma heats it, just like passing a current through any electrically conducting medium. A challenge is that the resistivity of the plasma falls as its temperature increases, and therefore, the resistive heating falls away. The result is that this so-called Ohmic heating can only heat the plasma to around 1 keV, significantly below fusion requirements. Additional heating is required; we refer to this as auxiliary heating. There are two broad forms of auxiliary heating – radio frequency (RF) waves and neutral beam injection (NBI).

In some ways, NBI seeks to replicate the effect of energetic alpha particles by firing beams of energetic particles into the plasma – typically the same species as the plasma ions. Of course, the injected particles must be neutral in order to pass through the magnetic field. This then raises the question of how to accelerate them to high energy. For this reason, one must start with ions which can then be accelerated to the required energy with an electric field. These ions are then passed through a gas to neutralise them before they enter the plasma. The energetic atoms are quickly ionised as they enter the plasma and heat it via collisions. One usually seeks to get the ions into the very core of the plasma to approach fusion conditions there. The hotter and/or more voluminous the plasma, the more energetic the neutral beam must be in order to penetrate to the core. On today's tokamaks, this typically requires a beam energy of up to 100keV, or so. Up to these energies, positive ions are relatively efficiently neutralised as they pass through the gas, picking up electrons in the process. However, future, larger tokamaks require higher energies, and then the neutralization efficiency of positive ions becomes very poor. Instead, negative ions must be accelerated, which do have a good neutralization efficiency at high energies. The challenge then becomes the efficiency of the production of the negative ions in the first place. Maximising this efficiency is an interesting research issue, which involves an understanding of complex low-temperature plasma physics.

RF waves will deposit their energy into the plasma when their frequency is tuned to a characteristic frequency of the plasma. There are a number of characteristic plasma frequencies, which then leads to different RF schemes. For example, in electron cyclotron resonance heating (ECRH), one matches the frequency of the RF waves to the frequency which electrons gyrate around magnetic field lines, which is typically in the microwave regime. This so-called cyclotron frequency is proportional to the magnetic field strength and as this varies through the plasma (falling off approximately inversely with R, the distance from the symmetry axis of the toroidal plasma), one can finely tune where heat is deposited by tuning the frequency of the microwaves. Other schemes match other characteristic plasma frequencies, such as the ion cyclotron frequency. The physics is complicated by so-called cut-offs in the plasma. These can be associated with natural plasma oscillations, which prevent an RF wave of a given frequency from passing through. It is instead reflected back out of the plasma. One such frequency is the plasma frequency, ω_{pe}, which is associated with a Langmuir wave – a wave that, in the simplest theoretical model, describes how electrons, if displaced from their associated ions, oscillate back and forth in a simple harmonic oscillator fashion with frequency ω_{pe}. The plasma frequency is proportional to the square root of the plasma density. Thus, by working in a plasma with a sufficiently high magnetic field compared to the plasma density, one can arrange for the microwaves to be damped at a given place in the plasma without encountering the cut-off frequency. Conversely, if the plasma density is too high, or the magnetic field too low, ECRH is not effective because of the cut-off. There is another interesting piece of physics that provides an alternative, however. Under certain conditions, which we cannot describe here, the electromagnetic ECRH wave can be converted to an electrostatic wave of the plasma called the electron Bernstein wave (EBW). This is not subject to the cut-off of the electromagnetic wave and so can propagate to the resonance position where its energy can be deposited. This EBW is not understood as well as the conventional ECRH wave, with outstanding research questions remaining around the physics of the conversion (including associated parametric instabilities), for example.

2.3.2   Confinement physics

The amount of heat that must be applied to bring the plasma to fusion conditions depends on how rapidly that heat is lost – in steady state, the heating scheme (including that from the alpha particles in a DT plasma) must provide power at the same rate it is lost from the plasma. Thus, plasma confinement is a key topic that underpins several physics aspects of tokamak power plant design.

In a perfect system, charged particles are tied to magnetic field lines and are therefore never lost in a toroidal geometry where those field lines lie in nested, closed toroidal surfaces. However, these charged particles gyrate around magnetic field lines in a helical path with a characteristic radius called the Larmor radius. Thus, if a particle gyrating around one magnetic field line collides with one gyrating around a different field line they can be scattered, and transfer energy across the magnetic field in the process. The result is a diffusive process where the step length of the diffusion is given by the Larmor radius, and the characteristic temporal step is the time between collisions. This is an irreducible, but extremely low, level of transport, called 'classical transport'. In a toroidal geometry, there is a more important transport mechanism associated with particle collisions, with similar physics to classical transport. We discussed earlier how variations in the magnetic field, together with the fact that magnetic field lines are curved, causes the charged particles to drift from their associated magnetic field lines. Toroidal symmetry means that particles will conserve the toroidal component of the canonical angular momentum as they travel. A result of this is that orbits are closed in the absence of any decorrelation mechanism (e.g., collisions). Most particles travel more-or-less freely along magnetic field lines, and the drifts only cause a small deviation from their associated field lines; these are called passing particles. However, there is a class of particles which have a component of their velocity perpendicular to the magnetic field line (i.e. the component responsible for the gyratory motion) rather larger than the component along it. These particles are constrained to the region of plasma with low magnetic field, on the outboard side at large R. As they travel along magnetic field lines, and as those field lines migrate into the higher magnetic field region at low R, this class of particles is reflected back into the low field region at a poloidal position known as the bounce point. However, they are all the time travelling under the influence of the drifts, which means they first move out from their associated flux surface in the top half of the plasma (for example) and then back towards the flux surface in the bottom half, to ultimately close the orbit. In the poloidal cross section, this orbit has the shape of a banana, so that is the name given to it (Figure 2.8). The width of the banana orbit – or the banana width – is usually much larger than the Larmor radius. Indeed, these so-called 'trapped particles' undergo a similar collisional diffusion to the classical process, but now the much larger banana width characterises the step length, resulting in significantly larger transport, which is further enhanced by the larger effective collision frequency associated with scattering trapped particles into passing ones. This process is called neoclassical transport. While neoclassical transport is much greater than the classical

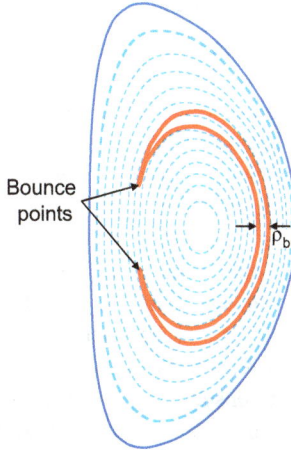

Figure 2.8   *The closed orbit of a trapped particle (full red curve) on the outboard side of a tokamak plasma, in the poloidal cross-section (the dashed curves denote magnetic flux surfaces). Note the bounce points, the banana shape of the orbit and the banana width, ρ_b. Trapped particles traverse along one leg towards the region of a higher magnetic field (on the left of the figure), then bounce back along the other leg.*

transport, if it were the only transport process in a tokamak, it would predict confinement times very much longer than those measured (e.g. illustrated in Figure 2.7). Indeed, the observed confinement time is about an order of magnitude (or more) shorter than that which would be predicted from the neoclassical transport theory. Something else is influencing the confinement time in tokamaks, and for a long time, this was not well understood. For this reason, the mechanism underlying the low observed confinement time was called 'anomalous transport'. We now know that mechanism is associated with plasma turbulence.

Plasma turbulence arises because of the complex interactions between the charged particles and the electromagnetic fields. So far, we have considered how plasma particles respond to the imposed magnetic field to create the confinement system. However, because the particles are charged, their relative movements can create electrostatic fields and a current density with an associated magnetic field. These in turn influence the motion of the charged particles. Under certain situations, feedback mechanisms between the particles and fields can drive small-scale fluctuations of the electromagnetic fields – these fluctuations are typically driven at the scale of the particle Larmor radii. These so-called 'micro-instabilities' can grow in amplitude, feeding off the free energy of the plasma gradients, such as the temperature or pressure gradient, depending on the mechanism of the instability. The result is a complex spectrum of fluctuations in the electromagnetic fields, driving turbulent flows in the electrons and ions all self-consistently coupled via Maxwell's equations. Coherent structures grow and decay within this turbulence, again driving a diffusive transport of heat and particles across the flux surfaces. The characteristic step length is now the size of the coherent structures, and the time step could be associated with their lifetime. While the concept is relatively simple, quantifying the resulting confinement time is extremely challenging. Part of the challenge arises from the physics of the interaction between particles and field fluctuations. Resonance behaviour occurs when the waves associated with the turbulent electromagnetic fields travel at the same speed as the particles. In the vicinity of a resonance, energy is readily transferred between the particles and the electromagnetic waves. This means that the nature of the turbulence drive depends on the details of the velocity distribution of the particles. The system then becomes six-dimensional: three spatial dimensions and three velocity dimensions. Furthermore, electron and ion responses to the fluctuating fields are coupled: the electrostatic potential adjusts to ensure quasi-neutrality (i.e. locally the positive ion charges and negative electron charges must balance) and Ampére's law (fluctuating currents generated by relative electron and ion flows must be consistent with the fluctuating magnetic field). Simulations of this complete, self-consistent system are intractable except in very simple geometries, but fortunately there is a rigorous theoretical model that can reduce the system – this is gyrokinetics.

Gyrokinetics makes use of the fact that the period of the gyromotion of electrons and ions around magnetic field lines is much shorter than the characteristic timescales of the turbulence; in other words, the turbulence changes very little during a single particle gyro-orbit. The theory of gyrokinetics exploits these disparate timescales to average over the particle gyro-motion, thus removing one of the velocity dimensions. The simulation problem remains a huge challenge, but further reductions are possible. One is the so-called 'flux-tube' approach [3]. This makes use of the fact that the characteristic turbulence length scale in the direction across magnetic field lines is comparable to the Larmor radius, which is typically a factor of 100 or so smaller than the tokamak system size. However, particles can stream more-or-less freely along magnetic field lines, meaning the structures in the turbulent field have very long length scales in that direction – comparable to the system size. One can therefore imagine analysing a narrow tube of plasma aligned with the magnetic field (there will be many small turbulent eddies across the tube), which is long enough to capture the long wavelength structures along magnetic field lines. However, the equilibrium parameters (e.g. the slowly evolving density, temperature, current density, magnetic field, etc.) do not vary significantly across the tube, although some might vary along it. One then makes the assumption that many statistically similar 'local' flux tubes can be combined to describe the turbulence in the vicinity of a given flux surface. Simulating the turbulence in such a flux tube is readily achievable with today's high performance computers. Nevertheless, challenges remain, such as quantifying the uncertainty and accuracy of the flux tube approach, and assessing the importance of complex multi-scale interactions between the fine scale turbulence and the macroscopic, slowly evolving equilibrium profiles.

If turbulence could be suppressed, the confinement time of tokamak plasmas would be dramatically increased. There are situations when this happens, the most robust of which is called the H-mode [4], or high confinement mode, which we now describe. Conversely, the turbulent, low confinement state is called the L-mode. As the heating power applied to an L-mode plasma is increased, the core plasma pressure gradually rises as expected. However, as the power is increased through a well-defined threshold, the confinement time suddenly increases, typically by a factor of two. This is called the L-H transition. A closer inspection of the density and temperature profiles in the H-mode state shows that a region of steep pressure gradient forms only in the last few centimetres of the plasma, just inside the separatrix flux surface shown in Figure 2.2. Measurements of the plasma fluctuations show that the turbulence is also suppressed in this edge plasma region, leading to a reduction in the heat and particle transport to enable the steep pressure gradient to form. Such a region of reduced transport is often called a transport barrier. Deeper into the plasma, beyond the transport barrier, the plasma remains turbulent with a modest pressure gradient. However, the whole of this central region is raised to a high pressure – as if it sits on a pressure pedestal created by the transport barrier. Indeed, the edge region is often called the pedestal.

The pedestal region of the H-mode is rich in physics, with numerous research questions remaining to be addressed [5]. First, how does it form? The widely accepted model is that strong flow shears are spontaneously formed in the pedestal region, and it is these that suppress the turbulence, tearing apart the plasma eddies to suppress the transport. Thus, in the L-mode, the edge pressure gradient is restricted to a low level because of fluctuations in the electrostatic field that drive the turbulence. As the flows develop, the turbulence is suppressed and the pressure gradient in the pedestal rises until it enters a regime where a second set of micro-instabilities can be excited to again drive turbulence and associated transport. These drive much stronger magnetic field fluctuations than the L-mode turbulence. An example of such micro-instability is called the kinetic ballooning mode (KBM), which tends to form ripples in the flux surfaces. Another is the micro-tearing mode, which tends to tear flux surfaces apart. These instabilities drive turbulent fluctuations that clamp the pressure gradient at a higher level, close to that required to trigger the instability. This process progressively widens the transport barrier, expanding the pedestal width. The combination of the pressure gradient clamped at the level required to trigger kinetic ballooning modes, and the pedestal width, determines the pedestal pressure – the pressure at the innermost edge of the pedestal. The pressure in the plasma core is observed to be approximately proportional to the pedestal pressure so, even though the pedestal is an extremely narrow annulus of plasma, it has a very big effect on the overall confinement. Thus, a first principles model of H-mode confinement must include a prediction for the pedestal pressure. We have briefly discussed our understanding of the pressure gradient arising from the electromagnetic turbulence, so to complete a model for the pressure pedestal requires us to address a second question of what determines the pedestal width? The combination of the pedestal width and its gradient then determines the pedestal pressure (with additional complications arising from plasma parameters at the separatrix, which we do not describe here.

To address the question of the pedestal width, we need to introduce another class of plasma instabilities. These are so-called magneto-hydrodynamic (MHD) instabilities, which are robust phenomena that involve a macroscopic movement of the plasma fluid. The resonances between the particle velocities and plasma wave propagation are less influential on the properties of these instabilities, which means that the detail of the velocity distribution of the particles is not as important as it is in the micro-instabilities that drive turbulence. MHD instabilities are therefore accurately described by treating the plasma as a highly (electrically) conducting fluid, which interacts with the magnetic field fluctuations. Thus, the MHD model describes how elements of the plasma flow under the influence of the pressure gradient and the electromagnetic forces. We focus on the ideal MHD model, whereby the plasma's electrical conductivity and viscosity are neglected [6].

Within the ideal MHD model, there are two sources of free energy that provide the main drives for instabilities: the pressure gradient drives ideal MHD ballooning modes (a more fundamental and robust form of the kinetic ballooning modes described earlier), and the current density gradient drives the kink family of modes. We have already discussed the high-pressure gradient that exists in the pedestal, but it is equally important to understand the current density. A tokamak plasma supports a current called the bootstrap current, which exists as a consequence of the trapped particles in the presence of a plasma pressure gradient. Indeed, the bootstrap current is proportional to the pressure gradient, and we have argued that this is large in the pedestal. Thus the bootstrap current density (and its gradient) is large in the pedestal. For this reason, the pedestal taps the free energy of both instability drives and is vulnerable to a class of instabilities called kink-ballooning or peeling-ballooning modes [7].

Peeling-ballooning modes have a radial extent that is rather larger than the micro-instabilities that drive the pedestal turbulence. Thus, when the pedestal is narrow, even though the pressure gradient is high, the peeling-ballooning modes will not fit in the pedestal region and they are not triggered. As the pedestal widens with a fixed pressure gradient constrained by the KBM, it eventually reaches a width that can contain the peeling-ballooning mode. This instability drives a large plasma eruption, collapsing the pressure gradient back down towards L-mode values to then re-build. Thus, we have a picture of a dynamic pedestal constantly building and collapsing under the influence of the peeling-ballooning mode. This then determines the maximal pedestal width that can be supported, providing a first-principles model of the pedestal height when combined with the gradient, which is constrained by the kinetic ballooning mode. This is the so-called 'EPED model', which has been widely validated by data from a range of tokamaks [8], although some discrepancies remain to be explained.

While the pedestal is clearly beneficial for tokamak plasma confinement, the penalty that comes with the plasma eruptions is a big one. Because the eruptions originate from the pedestal at the plasma edge, they are called Edge Localised Modes, or ELMs. These ELMs are filamentary in nature [9], ejecting hot plasma from the pedestal region towards material surfaces of the confinement vessel – up to around 10% of the thermal energy stored in the tokamak plasma. On today's tokamaks, ELMs provide a fascinating field of fundamental plasma physics research; on future tokamaks, they will cause excessive damage, eroding material surfaces with significant consequences for machine availability. Learning how to avoid or control ELMs is therefore essential, which we shall return to later.

2.3.3   Plasma exhaust

In steady state, all the power that goes into the plasma to achieve the fusion conditions – both the externally applied heating power and the fusion-produced alpha-power – must be exhausted. Managing this exhaust power is a major challenge. It is evident that the magnitude of the exhaust challenge is coupled to confinement: the worse the confinement, the higher the heating power that must be injected to achieve fusion conditions and the higher the heat that must be managed in the exhaust. This is a further example of why high confinement is so important for magnetic confinement fusion.

Plasma exhaust is usually managed in a tokamak through a specific geometry of magnetic fields called divertor geometry, as illustrated in Figure 2.9. As described earlier, within a special flux surface, called the separatrix, magnetic field lines lie on closed, nested magnetic surfaces, called flux surfaces. These provide good confinement and therefore support the plasma pressure gradient required to achieve fusion conditions at the plasma core. The separatrix is created by adjusting the currents in the poloidal field coils, appropriately positioned to create one or more nulls in the poloidal component of the magnetic field (a single null, denoted X-point, is illustrated in Figure 2.9). The flux surfaces outside the separatrix are 'open': they are cut by physical structures, called target plates. These target plates must carry the steady heat load arising from the exhaust, which for a fusion pilot

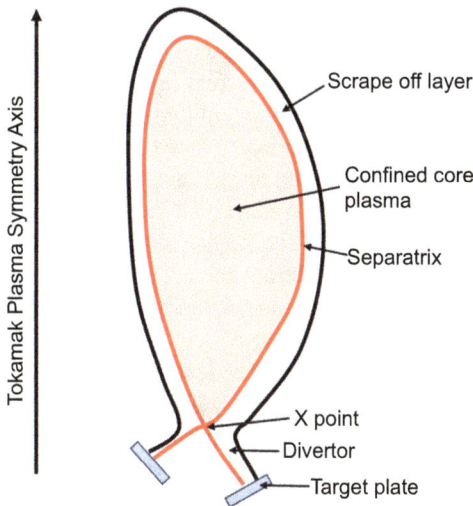

Figure 2.9   Sketch illustrating the key features of the divertor magnetic geometry

plant is at (or beyond!) material limits of around 10 MW m^{-2}, or more. The region of open flux surfaces which contains plasma is called the scrape-off layer. This acts as the tokamak's magnetic 'exhaust pipe', channeling particles and heat that diffuse from the core and across the separatrix into the divertor. This divertor region does not interface directly with the hot core plasma, and is a key region for managing the exhaust, as we now discuss.

There are two main mechanisms for exhausting the heat from the core plasma: radiation and convection. Radiation from hot gases can occur when electrons bound to ions are excited to higher energy levels by the thermal energy and then emit photons of characteristic frequencies as they relax back down to lower levels. However, the core deuterium and tritium are completely ionised, with electrons freed from their bound states within the atoms. Thus, this type of radiation is typically small in the core. It can, however, be enhanced significantly by introducing gas of higher atomic number – so-called 'impurities'. Indeed, the level of radiation increases strongly with the charge state so that the amount of radiation and its spatial distribution depend on the gas species introduced into the plasma. If significant high charge-state impurity ions manage to get into the very core of the plasma, then the radiation there provides a cooling effect that is detrimental to achieving fusion conditions. Indeed, even a fraction of a per cent of high-charge state tungsten ions in the plasma centre will radiate more power than provided to heat the plasma, resulting in a thermal collapse of the discharge. Thus, it is essential that high-charge state ions are screened from the core plasma. In the ITER tokamak [10], for example, tungsten is the material of choice for the divertor targets. While these are remote from the core plasma (Figure 2.9), it does raise an active research question of what conditions allow tungsten to migrate up the scrape-off layer to enter and pollute the core plasma.

While radiation in the hot, central region of the plasma core is undesirable as it is detrimental to achieving fusion conditions, radiation from the edge helps reduce the amount of heat entering the scrape-off layer and so lessens the impact on the divertor target plates. By adjusting the gas(es) injected into the plasma, one can in principle fine-tune where the radiation is emitted from. This radiation impacts the vessel wall relatively uniformly, helping to spread the exhaust power over a large surface area. Future pilot plants based on the tokamak concept are expected to rely heavily on this radiation to help reduce the power loads to the targets to a tolerable level.

Any power that is not radiated from the edge of the core plasma will be convected (by plasma turbulence) across the separatrix and into the scrape-off layer. It then flows very rapidly along the magnetic field lines until it arrives in the divertor, where it strikes the target plates. The heat load at these target plates is very high, but there is some uncertainty associated with predicting it in future tokamaks for the following reasons. Suppose the total power crossing the separatrix into the scrape-off layer is P_{sep}, and that the separatrix strikes the target plate at a major radius, R_t. The power per unit area striking the target plates – which must come below the maximum tolerable for the chosen target material and design – is then given by $P_{sep}/(2\pi R_t \Delta_t) \sin\delta$, where Δ_t is the width of the scrape-off layer at the target, and δ is the angle the magnetic field lines make with the target (Figure 2.10). The width Δ_u (Figure 2.10) arises as a result of the balance of the rapid flow of heat along magnetic field lines to the target and the slow diffusion of heat across the open flux surfaces. The disparate heat fluxes in these two directions results in a narrow scrape-off layer width. Quantifying the heat flux across the flux surfaces is especially difficult as it depends on the plasma turbulence in the scrape-off layer: the more turbulent the plasma there, the more rapid the transport across flux surfaces and the wider the scrape-off layer width, Δ_u, which then translates to a wide value at the target, Δ_t. Thus, while in the core plasma we seek regimes with low turbulence and corresponding good core confinement, we seek a turbulent regime for the scrape-off layer plasma to enhance Δ_t and reduce the power loads to the target plates. Considering that the pedestal region, just inside the separatrix and so interfacing with the scrape-off layer, is a region of low plasma turbulence, the challenge is evident. A key area of plasma physics research in magnetic confinement fusion is integrating a high-confinement core plasma with an acceptable exhaust plasma solution.

Let us consider some of the promising options available in designing an acceptable exhaust plasma. One is via the magnetic geometry. For example, bringing the divertor leg out to a larger major radius, R_t, lowers the heat flux to the target plates. Also, if one reduces the poloidal component of the magnetic field near the target plates compared to that at the tokamak plasma mid-plane, the flux surfaces will be more spread out in the divertor. This so-called flux expansion (equal to Δ_t/Δ_u – see Figure 2.10) spreads the heat over a larger area for a given value of Δ_u, thus reducing the flux towards an acceptable level for materials. Also, by

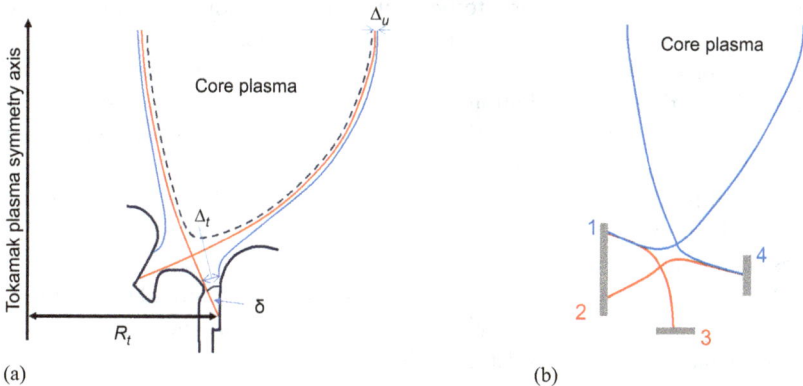

Figure 2.10   (a) The magnetic surfaces of the standard divertor configuration (blue is the outer edge of SOL, red is separatrix, black dashed is a core magnetic surface) illustrating features that can be exploited to reduce the heat flux to material surfaces: the major radius of the strike point, R_t, the flux expansion, Δ_t/Δ_u, and the strike point angle, δ. (b) Sketch of a snowflake divertor geometry showing an example with four strike points (labelled 1–4 at the grey target plates).

angling the target plates at an acute angle, δ, relative to the magnetic surfaces spreads the heat over a wider area. These features are illustrated in Figure 2.10(a), which shows the so-called single-null divertor (SND) geometry, with a single X-point, and two divertor legs: 'inner' at a small major radius and 'outer' at a large major radius. A natural question is whether the heat flux can be further reduced by employing a magnetic geometry with multiple X-points and/or divertor legs. A simple extension of the single null divertor is the double null divertor (DND), which typically employs an up-down symmetric equilibrium with X-points at the top and bottom of the plasma. Simple-mindedly, the power flux is halved, provided equal amounts of heat go to the upper and lower divertors. In fact, the situation is a little more complicated, both due to particle drifts and because this equal power flow requires both X-points to (almost) coincide with the same flux surface, which, in turn, requires a carefully balanced up-down symmetric equilibrium. This means developing a control system that can maintain the plasma's vertical position relative to the poloidal field coils with millimetre accuracy, which is especially difficult for highly elongated plasmas. Another way to create multiple divertor legs is by creating a higher order null in the poloidal magnetic field. An example, that has been trialed on the TCV tokamak, for example, is the so-called snowflake divertor, as illustrated in Figure 2.10(b) [11]. Again, a key question is whether the heat flux can be equally divided between the divertor legs.

One divertor geometry that captures many of the features discussed above is the so-called Super-X configuration, being explored on the MAST-U spherical tokamak (Figure 2.11) [12]. This employs a number of special poloidal field coils (the divertor coils) to pull the divertor leg out to large radius, while at the same time enhancing the flux expansion. As shown in Figure 2.11, this geometry allows the outer divertor region to be enclosed in a region far from the plasma. Impurity gas species can be introduced into this enclosed region, which then radiate the power over a wider area rather than restrict it to the narrow divertor strike point region. This radiation is further enhanced because of the large distance a field line travels before finally reaching the target plate. This is not so evident from the poloidal cross-section shown in Figure 2.11, as the large distance is a consequence of the field lines making very many toroidal revolutions (i.e. into the page of Figure 2.11) while they migrate over relatively small distances poloidally.

Our consideration of the Super-X divertor geometry provides a convenient segue from geometrical considerations to mitigate heat flux into another key concept: the detached divertor. A detached divertor corresponds to a situation whereby a gas layer exists in front of the target plate, so the divertor plasma plume interacts with this rather than directly with the material divertor plates. The result is that much of the power is radiated from this gas layer over a larger region. Detachment (i.e. the formation of the neutral gas layer) can occur when there is sufficient plasma density in the divertor region. It is a phenomenon that is influenced by the magnetic geometry of the divertor and the plasma density there, but also by the introduction of impurity gases to influence the radiation and therefore the temperature in front of the target plates. The interface between the scrape-off layer

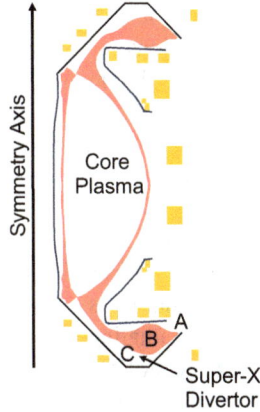

Figure 2.11   An illustration of some of the key features of the Super-X divertor configuration of the MAST-U spherical tokamak, showing (A) the ability to pull the outer divertor leg out to a large radius, R_t, (B) the possibility for larger flux expansion, (C) the closed divertor region into which impurities can be injected to radiate power over a wider area. Black lines denote material surfaces; copper-coloured boxes are the circular poloidal field coil cross-sections, and the pink area represents the scrape-off layer plasma. The whole system is symmetric about the axis shown.

plasma of the divertor leg and the gas layer is called the 'detachment front', and a key challenge is its control. Varying plasma conditions (e.g. divertor plasma density) influence the position of this detachment front, which is important–if the front extends up the divertor leg as far as the X-point, the core plasma confinement can be degraded. Indeed, it can sometimes result in a rapid, irreversible thermal collapse accompanied by an influx of impurities into the core and leading to a sudden termination of the discharge. Thus, detachment and the control of the front are thought to be key for the operation of plasma exhaust systems in future tokamaks, and therefore quantifying and developing the control requirements for the detachment front is a key research activity in tokamak physics [13].

We close this section on tokamak plasma physics with a discussion of a topic that is one of the biggest challenges facing this technology: disruptions. A disruption is a dramatic and sudden loss of the plasma, resulting in an unplanned termination of the discharge. The consequences are far greater than the inconvenience of a shorter-than-planned plasma pulse, as disruptions on future, larger tokamaks could cause terminal damage to the machine if unmitigated. There are three key damaging phenomena associated with a disruption. First, the loss of control of the plasma results in a rapid cooling of it (thermal quench), imparting huge thermal loads on the plasma-facing components inside the tokamak vessel. Second, the plasma current that is provided to support the poloidal component of the magnetic field decays rapidly (current quench); this drives large, so-called, halo currents in the vessel structure. These currents will be measured in MA for future tokamaks like ITER and fusion power plant designs which, when crossed with the confining toroidal component of the magnetic field (a few Tesla), results in mega-Newtons of electromagnetic force on the internal tokamak components and structures. Third, the decay of the current drives an electro-motive force in the plasma, which slows the decay of that current (Faraday's Law). The resulting electric field accelerates the very light electrons towards relativistic speeds, experiencing very little friction against the background plasma. These so-called 'runaway electrons' are so energetic that if unchecked they cause significant damage. These detrimental impacts can be mitigated by injecting large numbers of neutral particles into the plasma before the disruption develops, thus radiating away much of the energy. In ITER, these are envisioned to be large pellets of frozen material, each the size of a wine bottle cork, which are fired along a guide tube and shattered before entering the plasma. This 'shattered pellet injector' (SPI) is a relatively new technology, with recent experiments on JET seeking to identify a reliable design for ITER [14]. There remains much to learn, such as the optimal design of the injection system and pellets, as well as a reliable disruption detection technique to provide the trigger for the SPI.

2.4   Tokamak materials and technologies

While the main aim of this chapter is to describe some of the key plasma physics research challenges associated with magnetic confinement fusion power – and tokamaks in particular – there are a number of ancillary technologies that must also be developed. Furthermore, these technologies are often coupled with plasma physics challenges, so it is appropriate to briefly consider them here.

2.4.1   Materials science

In the previous section, we considered some of the conditions that materials must withstand from a plasma perspective, such as the thermal loads they need to be designed to tolerate, including both steady-state situations (e.g. the divertor heat flux to material target plates) as well as transients, such as the ELMs that we described. While the heat loads certainly do pose a challenge to materials, one can (and does) develop experimental rigs to test and develop prototype components in environments representative of the thermal loads in future power plants. However, it is much more difficult to recreate the fusion neutron environment that materials will need to withstand.

The neutrons produced in a DT fusion reaction have an order of magnitude higher energy than neutrons produced in a typical fission reaction. They fly out from the plasma and interact with the materials that constitute the various plasma-facing components, typically penetrating a metre or so through them. As they travel through the components, they slow down. Thus, it is only those components that are closest to the plasma that experiences the most energetic neutrons, while those that are shielded behind other components will experience slower neutrons more characteristic of a fission spectrum. This is key because the damage mechanisms differ for different energy neutrons. One way to visualise this is to consider a fast neutron striking an atom in the crystal lattice of a material structure. A sufficiently high-energy collision will displace that atom, so that now we have both our initial neutron and an energetic ion travelling through the material, both of which can strike other ions at other lattice sites, displacing them and so on to create an avalanche of particles crashing through the material, all from the initial neutron. This seems devastating, but as the particles' energies become absorbed and the avalanche comes to an end, the crystal structure has a natural self-healing tendency, and ions adjust their positions to re-establish the crystal lattice, thus largely annealing the initial damage. Nevertheless, the healing is not perfect, with some atoms being out of line with the crystal structure, and gaps existing at some locations in the lattice. Furthermore, these 'defects' will migrate through the material and typically congregate in the vicinity of grain boundaries (a material sample is not typically one large single uniform crystal, but is normally made up of 'grains', each of which has the characteristic lattice structure of the material). In addition, nuclear reactions will occur between the neutrons and the elements that the material is composed of, causing transmutations to different elements. These elements themselves are often unstable and susceptible to radioactive decay. This creates a radioactive waste legacy (which we return to later) but can also create helium as a result of alpha decay, for example. This helium can accumulate at the defects mentioned earlier, forming helium bubbles. Not surprisingly, all of these mechanisms modify the material's structural properties significantly. There is also very significant swelling (e.g. associated with the helium bubbles) so that components will physically change in size over time as they are exposed to more and more neutrons from the fusion reactions.

The engineering design of a demonstration fusion plant must take account of this neutron-induced damage. However, it is apparent from the description of the mechanisms above that the extent and type of damage will be highly dependent on both the neutron energy spectrum and the elements that the materials are made from. Tests of materials' structural properties following irradiation are clearly important. Ideally, this should include the neutron irradiation impact on joining methods (e.g. welds, bolts, etc.) as well as on the pure material itself. In the absence of an energetic fusion neutron source with sufficient fluence, a range of alternatives are employed. For example, materials are exposed to neutrons from fission test reactors and then tested. These are often tests on small samples to determine the impact on structural properties, which then have to be scaled up to the full component size. As mentioned earlier, these fission neutrons have lower energy than DT fusion neutrons and therefore have different damage mechanisms. Nevertheless, the data is valuable for understanding the properties of components that are shielded to some extent from the plasma, for which the neutrons will have slowed significantly. To some extent, the data can also be extrapolated to understand materials properties following high energy neutron irradiation, especially when combined with modelling to provide a physics-based predictive capability. Additional data to substantiate models can also be taken from ion irradiation experiments. High energy beams of ions can readily be generated by accelerating them through an electric field. They can then be used as a proxy for the neutron irradiation. However, their charge means that they have a much stronger interaction with the ions in the material lattice than a neutron has, and so they are stopped within the first few microns, whereas neutrons will typically penetrate up to a metre, or so. While the initial interaction is different, the ion beam can nevertheless stimulate the cascade process that we discussed above, and therefore can provide valuable information to inform and validate models of the cascade and subsequent annealing.

It is apparent from the above discussion that the experimental conditions that we can currently create fall some way short of those that materials and components will experience in a future fusion pilot plant. This motivates the construction of a new facility to create a high fluence of neutrons with an energy spectrum that is more representative of that to be experienced by the fusion components near the plasma. Europe is most advanced in this activity, with the construction of the billion-Euro-scale IFMIF-DONES facility recently initiated in Grenada [15]. IFMIF-DONES is designed to accelerate a beam of deuterons over 100 m to 40 MeV, whereupon it strikes a lithium target to produce energetic neutrons through stripping reactions. These neutrons will then impact upon up to about 1000 small material samples contained within a 0.5 l test volume [16]. Scientific challenges will remain, requiring extrapolation of the data to multi-material, multi-physics systems that include interfaces and joints – direct testing of these components would require a neutron source capable of irradiating on the cubic metre scale and as yet such facility designs are only developed to the conceptual level.

As well as influencing the structural integrity and functional performance of materials, neutron irradiation also influences the waste. Long-lived, high-level radioactive waste is avoided in a fusion plant, and this reduces the need for complex storage facilities and the controversy over the environmental legacy. Nevertheless, the need to manage low-level waste is inevitable, but this is relatively straightforward to store and typically has a lifetime of a hundred years or so. Furthermore, this waste legacy can be further minimised by careful design of materials. Consider steel, for example. Steel is dominated by iron, of course, but the impact of neutrons on natural iron isotopes is relatively benign (any radioactive isotopes that are generated via the neutron irradiation decay in a matter of minutes). However, steels have a mix of other elements, and it is this mix that gives those steels their distinct structural and functional properties. While these additional elements are introduced at a very low level (around 1%, or less), it is these that can be converted to radioactive isotopes by neutron irradiation. The aim, therefore, is to seek 'low activation' steels, such as EUROFER, which employ element isotopes that minimise the waste and yet retain the material's required structural properties. A take-away message is that the activated waste from a fusion plant is low-level and relatively short-lived, and because it largely arises because of the materials the plant is made out of and not from the fusion reaction itself, there is significant scope for materials science and technology to further reduce the waste legacy.

2.4.2   Tritium breeding

One of the biggest unproven technologies for DT fusion is the breeding of tritium. This requires an arrangement of blanket modules to be installed around the plasma, aiming to capture as many of the fusion neutrons as possible. Each tritium nucleus that is burnt in a fusion reaction produces one neutron, so, provided that this neutron reacts with a lithium-ion in the blanket to produce a tritium ion, the amount of tritium is conserved. Of course, this is not sufficient for two reasons: (1) not every neutron will be captured by a lithium-ion (many will be lost to the material structures of the vessel, for example), and (2) one needs to generate more tritium than is burnt in order to provide fuel to start other fusion devices. This means that the breeding blankets must multiply the neutrons (e.g. using beryllium or lead), moderate their energy to optimise the cross section with lithium and employ an optimal mix of ^{6}Li and ^{7}Li isotopes, noting that the dominant naturally occurring isotope is ^{7}Li (about 95%). When combined with a need to extract the heat via an appropriate coolant (and convert it to electricity) and to extract the tritium produced, one can see that the blankets are complex, multi-physics components. Given that their performance is essential to the sustained operation of the power plant, it is important that appropriate facilities are developed to test and optimise prospective designs. Such facilities should not only test the ability to breed tritium but must also demonstrate effective cooling of the first wall facing the plasma, guiding the choice of coolant as well as innovative technologies to maximise the heat transfer from the plasma-facing surfaces to the coolant.

2.4.3   Tritium cycle and helium ash

Managing the tritium in a fusion power plant goes beyond its production in the blanket. To fuel a tokamak plasma, for example, tritium and deuterium will likely be formed into frozen pellets (much smaller than the pellets used for disruption mitigation) and injected into the plasma at high velocity. These pellets rapidly ablate, so a high velocity is required to ensure they reach the plasma core. They are then ionised to become part of the burning plasma. The tritium will diffuse through the confining magnetic field to the plasma edge in a time related to the confinement time, which is typically of the order of seconds in a tokamak, when it is then exhausted from the plasma. Of course, the exhaust gas will not only be tritium – it will also include deuterium, any impurity ions that have got into the plasma, and the helium that is produced in the fusion reaction. It is important to create plasma conditions such that the fusion-produced helium ions will give up their energy to heat the deuterium–tritium mix plasma before they are expelled. Once they have lost their energy, the helium ions serve no purpose and only dilute the DT fuel, diminishing the fusion power output – it is then referred to as 'helium ash'. It is important to extract the helium ash in the exhaust at a certain minimum rate, which means that the particle confinement time should not be too great in an effective fusion device. Thus, the exhaust requires a significant level of processing, removing the helium 'ash' and extracting out pure tritium to be reintroduced into the plasma. Each tritium ion will typically go around this loop tens of times before finally being burnt. Developing this closed fuel cycle is a key technology to be established, involving not only the extraction and purification of the tritium, but also ensuring a choice of materials that minimises the amount of tritium held in the structure. This is important for minimising the tritium inventory of the plant, which otherwise has implications for the licensing and cost (tritium being a valuable commodity). Tritium also complicates waste management if it is absorbed into the structure or coolants, especially water. Developing effective tritium permeation barriers is an important strategy for minimizing the amount of tritiated waste that needs to be managed. Detritiation technologies are also key to further reduce the waste legacy from fusion energy.

2.4.4   Enabling technologies

There are a number of technologies required to enable the fusion conditions, and the availability of the plant. We cannot go into detail here, but we will mention a few key issues associated with each. Different approaches to fusion have different enabling technologies, but we will continue to use the tokamak as an example.

Key to the tokamak are the magnets to provide the magnetic confinement. In modern tokamaks, these are typically superconducting, employing the well-developed technology of standard, low-temperature superconductors. The push to higher magnetic field to further improve confinement and seek more compact solutions for a tokamak pilot plant with lower capital cost is driving substantial research into high-temperature superconductors. These are beneficial because they can provide significantly higher magnetic field than low-temperature superconductors, while maintaining their superconducting properties. Nevertheless, research challenges remain, such as their behaviour in a neutron environment and the feasibility of joints that can be broken and re-assembled for maintenance. There are also significant engineering challenges associated with the very high electromagnetic forces that are generated. Nevertheless, high-temperature superconductors could open up new pathways to commercial fusion energy and a number of organisations – private and public – are exploring their potential.

Heating and current drive schemes are required to bring the plasma to fusion conditions and also to provide a steady state current when long pulse operation is required. Neutral beam injection is a robust scheme and is a major workhorse for most of the tokamaks in operation today. However, there are challenges for future tokamaks, such as the need to efficiently produce negative ion beam sources in an environmentally friendly way, and the large ports required for these systems. These ports take up space that would otherwise be used for tritium breeding and therefore threaten the requirement for achieving a tritium breeding ratio in excess of unity. The result is that radiofrequency waves are the most likely heating scheme for a fusion pilot plant. Amongst these, microwaves are attractive as they do not require a large antenna near the plasma. Significant research is required to further improve the efficiency of the gyrotrons that are used to generate microwaves and to push them towards reliable steady state operation.

The first fusion pilot plants are bound to be unreliable. We will learn much from their first few years of operation, which will then feed into robust designs for the commercial fleet of fusion power plants to follow. However, it is important that pilot plant designers acknowledge and quantify this uncertainty in reliability and ensure an effective remote maintenance and inspection scheme informs their designs from the outset. The harsh environment of a tokamak will put additional demands on remote maintenance and inspection systems which will require new approaches to ensure availability is maximised.

2.5 Conclusion – the pathway to fusion delivery

In this chapter, we have explored the research challenges that face the development of fusion energy as a commercially viable prospect. We have focused mainly on one particular example – the tokamak – to provide a concrete use case for our discussion. While this is presently the most advanced approach to commercial fusion power, alternatives – such as the stellarator or inertial confinement schemes – have additional advantages, and it remains important to keep a portfolio of alternatives on the table at this stage.

A frequent – and important – question is how long must we wait to see the first fusion power delivered to the grid? Given the urgency driven by climate change, and the need for a sustainable, safe, secure energy supply to power our society and industries, this is not the right way to pose the question. Many governments around the world have greenhouse gas emission ambitions that target 2050 or earlier. Thus, one must turn this question about delivery timescale around and ask it as a research question: how do we deliver fusion power to the grid by 2050?

A requirement to meet this timescale is to have a demonstration fusion pilot plant operational by around 2040. Such a device should demonstrate net power to the grid and a closed fuel cycle. However, while it must point the way to commercially viable fusion power, it does not need to be commercially viable itself. Our experience of building ITER is invaluable for the pilot plant design and construction. Although ITER is designed as an experimental facility and will not deliver power to the grid, it is the world's first industrial-scale fusion plant and provides many lessons about integrating fusion systems at scale, whatever the fusion pilot plant concept. However, it will not produce a neutron fluence representative of a fusion pilot plant, and therefore, while it will be invaluable for developing optimised burning plasma scenarios and some technologies, it will not provide an adequate testing environment for many of the fusion nuclear science issues that a fusion pilot plant will face.

A fundamental issue is that we need to develop and test fusion components in a representative fusion environment including thermal, electromagnetic and neutron loads. The only way to create this integrated set of conditions is with a fusion reactor. Let us start by examining whether or not we can use the fusion pilot plant itself as its own test bed. Figure 2.12 shows a generic DT fusion pilot plant system. Note the close coupling between the different subsystems, and the lack of external inputs (actuators) to the system for varying the fusion conditions. Heating (and current drive for a tokamak) provides some level of external influence, but in the

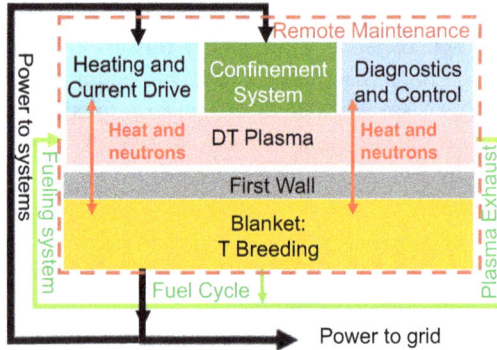

Figure 2.12   A generic fusion pilot plant system, showing the closely coupled sub-systems. While essential as a demonstration of the viability of fusion power, there are relatively few actuators available to vary the environmental conditions to provide a flexible experimental facility.

pilot plant, this will be dominated by the alpha heating process that is internal to the plasma. Indeed, the need to demonstrate net power to the grid requires a design that minimises external heating power. It also drives a minimum size to the fusion pilot plant, pushing one towards GW-scale fusion power, which then requires a significant amount of tritium. Thus, the pilot plant operation will be crucially dependent on the ability to produce more tritium in the blanket than is burnt in the plasma core. Given the blanket is one of the systems we will be seeking to test, this creates a high technical risk: if the blanket does not perform, it will not even be possible to create the fusion conditions required to test and optimise that blanket effectively, which could then lead to long delays to the final demonstration of net fusion power to the grid. In summary, the fusion pilot plant is an essential demonstration device on the pathway to fusion commercialization, but it is not sufficiently flexible to serve as an experimental facility. Without an experimental capability in addition to the fusion pilot plant, there is a very significant risk that the 2050 target for fusion demonstration will be missed.

If we relax some of the requirements compared to those of the fusion pilot plant, can we envisage a single experimental facility that can deliver an integrated fusion environment? Two of the biggest drivers for the fusion pilot plant are demonstration of net electricity and, related to that, a closed fuel cycle without reliance on an external supply of tritium (beyond the initial start-up charge). Relaxing the net electricity requirement opens up options for lower cost, smaller fusion devices with lower fusion power and, therefore, reduced tritium consumption. Figure 2.13 shows an adaptation of the fusion pilot plant schematic of Figure 2.12, illustrating how such a scenario might look. Power is now drawn from the grid to operate the facility (while still benefiting from the fusion alpha heating power to supplement external heating power), and it is assumed that the tritium consumption is sufficiently low that it can be obtained from commercial sources, rather than relying on the breeding blanket technology. Considering the

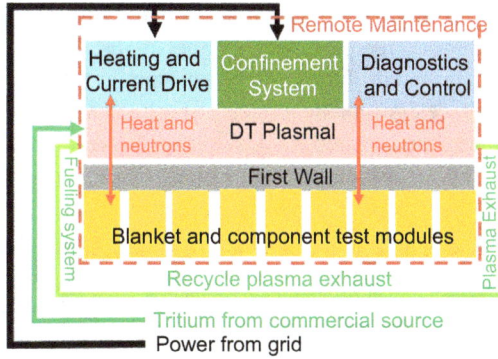

Figure 2.13   An option for an integrated fusion environment test facility, relaxing the requirement for producing net power to the grid, and not reliant on a closed fuel cycle, but instead buying tritium from a commercial source.

specific example of a tokamak, the need to shield magnets from the fusion neutrons pushes one towards a minimum major radius of around $R = 2.5$ m, which then requires a fusion power of 100–200 MW, or so, to provide sufficient neutrons for an adequate testing regime – substantially less than the GW-scale fusion pilot plant power anticipated, and therefore requiring significantly less tritium. Indeed, availability of such a tokamak of about 10% would translate to a tritium consumption of about 1 kg per year, and an average neutron wall loading over the first wall approaching 1 MW m^{-2}, sufficient to start exploring how fusion subsystems perform in integrated fusion conditions. The fact that we do not need to breed tritium to operate this device means we can replace the tritium breeding blankets with multiple material/component test modules arranged around the plasma (Figure 2.13), which can be carefully controlled to enable rigorous experiments to be performed. More advanced plasma scenario options, combined with some tritium breeding capability (e.g. during a Phase 2 of operations, after we have learnt how to optimise the blankets in Phase 1) would lead to fusion conditions approaching those of the fusion pilot plant, but in a smaller, more flexible device.

Such a single integrated test facility would provide a key experimental capability, with sufficient flexibility to explore a number of subsystem design options and operating scenarios in an integrated fusion environment. It is a simpler system than the fusion pilot plant, but it is nevertheless a nuclear facility that is expensive and would require careful design and construction. An ambitious timeline could see it operating just after 2035, which could provide some input to construction of the fusion pilot plant and its subsequent operation. However, the main benefit will be as a key fusion *technology experimental* facility operating alongside the ITER (largely) *physics experiment*, with the fusion pilot plant operating in parallel as a commercial *demonstrator*.

This powerful trio of facilities could support the timely delivery of the first fusion reactors of the commercial fleet by the 2050s, but the cost of such an approach would be significant. Furthermore, substantial technical risk remains for the fusion pilot plant design: an integrated experimental technology test facility such as that described above would be difficult to deliver in time to influence the pilot plant design, and wrong choices here could lead to delays in its ability to demonstrate the commercial pathway. To mitigate this risk, we need fusion technology data within the 2020s to make informed decisions about fusion pilot plant design. This can be achieved in principle by relaxing the condition of testing in a fully integrated facility and instead breaking the system down into a number of (semi-)independent test facilities that represent the different components of the fusion pilot plant: a neutron source (such as IFMIF-DONES in the EU or the Fusion Prototypical Neutron Source being explored in the US) to test material samples; a blanket component test facility to explore the thermo-hydraulic properties of different design options (perhaps with magnetic field); a fuel cycle test facility to ensure efficient processing of the tritium fuel; a high heat flux facility to develop and test aspects of the first wall, and a suite of facilities to prototype and test a range of enabling technologies (inertial fusion drivers, magnets, heating systems, remote maintenance, etc). Such a suite of facilities would help fill some knowledge gaps, but very significant ones will remain related to: (1) the inability to create the fully integrated fusion environment in a single test facility; (2) the inability to explore the interfaces between different subsystems; and (3) related to these, how to perform the very significant extrapolation to the fusion pilot plant. An exciting opportunity is to explore the emerging capability of advanced computing, augmenting the physical test facilities with virtual ones. Whether this is a viable option for reducing (and quantifying) the technical risk associated with delivering the fusion pilot plant by 2040 is itself a research question. However, if such an approach can be developed and proven it will have applications in multiple other sectors that seek to accelerate the early delivery of a 'future technology'.

It is worth considering a particular approach to provide a specific example of the power of integrating the real and virtual worlds. This is illustrated by Figure 2.14. Each hexagon in the figure represents two closely coupled entities: the physical test facility (or collection of test facilities) to address a particular subsystem and a high fidelity 'exact' virtual representation of it provided by advanced engineering simulation. This close coupling between the two entities provides a convenient validation capability, giving confidence in the simulation at least within the conditions that the physical testbed can access. Furthermore, provided that the virtual representation is built upon rigorous physical science and engineering foundations, one can simulate the performance of the system in conditions beyond those accessible to a single physical test rig, closer to those of the pilot plant. This then addresses one key requirement: the ability to extrapolate the performance of the test rig towards fusion pilot plant conditions. However, such high-fidelity simulations will require substantial computing resource, and it is unlikely they will be able to provide the basis for exploring the challenges associated with integrating the various subsystems into the full fusion pilot plant design. To do this, one needs

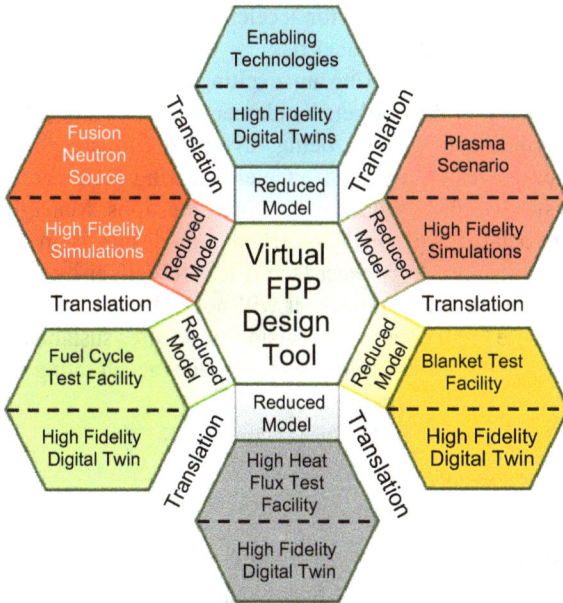

Figure 2.14   How physical test rigs and high fidelity engineering simulations might be combined, building on the digital twin philosophy, to inform reduced fidelity models that can be integrated into a virtual fusion pilot plant design tool, capturing the data from fundamental experiments and translating them into the integrated design of fusion pilot plant components

to develop lower fidelity models that can be simulated much faster and are amenable to combination into a fully integrated simulation of the fusion pilot plant design: a virtual fusion pilot plant. Such reduced models will be informed and validated by data from both the physical and high-fidelity virtual test beds, using the latter to provide simulation data relevant to the fusion pilot plant. Furthermore, the high-fidelity simulations will illustrate which mechanisms are important to capture in the reduced models, informing their construction. Finally, and importantly, the high fidelity simulations can be used to quantify the uncertainty in the reduced fidelity modules, which can then be propagated through the full virtual pilot plant design to inform the confidence level with which it will meet its fusion targets (e.g. tritium breeding and power to the grid). As the fusion pilot plant, ITER DT plasmas and the integrated fusion technology test facility come online from the mid-2030s, they would provide the final data to optimise the ultimate virtual design tool and so rapidly establish an efficient, commercially viable fleet of fusion power plants from the middle of this century. The challenge associated with this approach should not be under-estimated and the resource required is significant, but it is no greater than the challenge associated with building an actual fusion pilot plant. Furthermore, if the approach is successful, it could revolutionise how we approach engineering design and construction for accelerating a range of future technologies, and managing the associated technical risk.

To conclude, fusion is entering an exciting new era, working aggressively towards early delivery to the market. Fundamental research questions remain, but to meet the demanding timescales being driven by climate change and energy insecurity, we need to learn how to advance engineering design and construction in parallel with answering these final fundamental questions. Managing the inherent technical risk associated with such uncertainty in design is in itself a research challenge and one fusion energy must face if it is to deliver by 2050. However, if we can identify a successful approach, it will not only address one of the biggest challenges facing the human race – provision of secure, sustainable energy – but also find applications in multiple other sectors where accelerating the development of future technologies is important.

References

[1]   H. Abu-Shawareb, R. Acree, P. Adams, *et al.* 'Achievement of target gain larger than unity in an inertial fusion experiment', *Physical Review Letters*. 2024; 132: 065102

[2]   M. Sparkes, 'UK nuclear fusion reactor sets new world record for energy output ', *New Scientist*. https://www.newscientist.com/article/2415909-uk-nuclear-fusion-reactor-sets-new-world-record-for-energy-output/ 2024

[3]   M.A. Beer, S.C. Cowley and G.W. Hammett 'Field-aligned coordinates for nonlinear simulations of tokamak turbulence'. *Physics of Plasmas*. 1995; 2 (7): 2687–2700

[4]   F. Wagner 'A quarter-century of H-mode studies'. *Plasma Physics and Controlled Fusion*. 2007; 49: B1–B33

[5]   R.J. Groebner and S. Saarelma 'Elements of the H-mode pedestal structure'. *Plasma Physics and Controlled Fusion*. 2023; 65: 073001

[6]   J.P. Freidberg *Ideal MHD*. Cambridge:Cambridge University Press; 2024

[7]   J.W. Connor, R.J. Hastie, H.R. Wilson and R.L. Miller 'Magnetohydrodynamic stability of tokamak edge plasmas'. *Physics of Plasmas*. 1998; 5: 2687–2700

[8]   P.B. Snyder, R.J. Groebner, J.W. Hughes, *et al.* 'A first-principles predictive model of the pedestal height and width' *Nuclear Fusion*. 2011; 51: 103016

[9]   C. Ham, A. Kirk, S. Pamela and H.R. Wilson 'Filamentary plasma eruptions and their control on the route to fusion energy' *Nature Reviews Physics*. 2020; 2: 159–167

[10] K. Ikeda 'Progress in the ITER Physics Basis'. *Nuclear Fusion*. 2007; 47: E01

[11] B. Labit, G.P. Canal, N. Christen, *et al.* 'Experimental studies of the snowflake divertor in TCV'. *Nuclear Materials and Energy*. 2017; 12: 1015–1019

[12] J.R. Harrison, C. Bowman, J.G. Clark, *et al.* 'Benefits of the Super-X divertor configuration for scenario integration on MAST Upgrade'. *Plasma Physics and Controlled Fusion*. 2024; 66: 065019

[13]   B. Lipschultz, F.I. Parra and I.H. Hutchinson 'Sensitivity of detachment extent to magnetic configuration and external parameters'. *Nuclear Fusion.* 2016; 56: 056007

[14]   S. Jachmich, U. Kruezi, M. Lehnen, *et al.* 'Shattered pellet injection experiments at JET in support of the ITER disruption mitigation system design'. *Nuclear Fusion.* 2021; 62: 026012

[15]   W. Królas, A. Ibarra, F. Arbeiter, *et al.* 'The IFMIF-DONES fusion oriented neutron source: evolution of the design'. *Nuclear Fusion.* 2021; 61: 125002

[16]   A. Ibarra, F. Arbeiter, D. Bernardi, *et al.* 'The IFMIF-DONES project: preliminary engineering design'. *Nuclear Fusion.* 2018; 58: 105002

Chapter 3

The role of metamaterials in future electromagnetic technologies

Rebecca Seviour[1] and Jonathan Gratus[2,3]

3.1 Introduction

Since the start of the twenty-first century, research into metamaterials has grown at a frantic pace, with researchers proposing new materials with increasingly more exotic properties and proposed applications. In this chapter, we aim to present an overview of the field, advantages, and exotic properties these materials can offer, and emerging application areas for these materials. The potential applications for metamaterials are diverse and promising. They have been proposed as candidates for optical filtering, medical devices, remote aerospace operations, sensor detectors, solar power management, crowd control, radomes, antenna lenses, and many more. Ultimately, this is a very broad and rich area, which, in this chapter, we can only hope to give the reader an overview of, with the hopes this inspires the reader to explore aspects presented in greater depth.

From optical to RF wavelengths, the propagation and interaction of electromagnetic (EM) waves are governed by Maxwell's equations. Where at the microscopic level, these fields couple to the material they are incident on and propagate through by inducing polarization and magnetization of the atoms in the material. In Maxwell's equations, the coupling between materials and the electric/magnetic fields is described by two abstract material parameters: the permittivity ε and the permeability μ.

For over a hundred years, scientists have synthesized molecular materials to give tailored ε and μ, determined by the movement of the light-mass negatively charged electrons surrounding the relatively large-mass positively charged nucleus of atoms in response to an EM wave forming a dipole. However, this interaction can only offer a limited range of ε and μ due to the fundamental properties (charge, mass) and the chemical bonds formed by the atoms of the material. This limitation has led scientists and engineers to consider a range of artificial composite structures with periodic sub-wavelength functional inclusions. Although these inclusions are

[1]School of Computing and Engineering Ion Beam Centre, University of Huddersfield, UK
[2]Physics Department, University of Lancaster, UK
[3]The Cockcroft Institute of Accelerator Science and Technology, Sci-Tech Daresbury, Warrington, UK

many orders of magnitude larger than the molecules of the constitutive materials, they are still much smaller than the incident EM wavelength, meaning these inclusions respond no differently than giant molecules with very large polarizability. Enabling the interactions between wave and the collective structures to be described in terms of the 'homogenized' abstracted bulk material parameters permittivity and permeability. Treating the collective periodic structures as an 'effective' material. This approach in theory allows engineers and scientists to fabricate artificial materials with specific electromagnetic properties, most notable of which is the creation of materials with simultaneously negative ε and μ. There are of course still restrictions on engineered properties; for example, it is impossible to engineer a material where the group velocity of an electromagnetic EM wave is greater than the speed of light in a vacuum.

3.2   History

The word metamaterial entered the lexicon around 1999, attributed to Roger Walser, defined as '...macroscopic composites having man-made, three-dimensional, periodic cellular architecture designed to produce an optimized combination, not available in nature, of two or more responses to specific excitation', to refer to certain types of effective media. Although the history of effective media predates the introduction of the term metamaterial by more than 100 years, it builds on the theoretical framework developed in the nineteenth century by Mossitti [1] and Clausius [2]. Clausius–Mossotti homogenization expresses a material's relative permittivity in terms of the atomic polarizability of its constituent atoms, molecules, or a homogeneous mixtures. For example, if a system of particles is small enough, then its reaction to an EM wave is identical to that of a collection of molecules with high polarizability. In general, when the inhomogeneities are small compared to the incident wavelength, the system appears homogeneous to the wave. Homogenization enables us to describe the EM behaviour of a heterogeneous system by evaluating the effective permittivity and permeability of an equivalent macroscopically homogeneous medium. The effective permittivity and permeability of the bulk are found using sub-wavelength elements of the system. This approach is the basis for many 'effective media' theories, such as the Maxwell-Garnett [3] and the Bruggeman [4] approach. Each approach is based on slightly different assumptions about the topology and material properties of the constituent materials. The Maxwell-Garnett approach implies that the inclusions are well-defined spheres sparsely scattered across the host medium. The Bruggeman approach is an essential percolation approach, where the two mediums are equally mixed. These examples highlight a key point about effective media theories. As effective-permittivity/permeability are averaged differently in each model, different effective-media theories cannot be directly compared, even when the same sub-wavelength configuration is used. There are too many effective media theories to discuss here; if the reader is interested in this area, then we highly recommend the comprehensive reviews by Lakhtakia [5] and Belov [6], which also discuss metamaterial homogenization.

3.3 Constitutive relations

As stated in the Introduction, metamaterials are sub-wavelength structures that are engineered so that the material has desired constitutive relations. Since they have the potential to enable novel constitutive relations that do not exist in nature, we consider here the questions of what exactly constitutive relations are and what the range of possibilities opened up by the use of metamaterials.

In a medium, one uses Maxwell's macroscopic equations:

$$\nabla \cdot \mathbf{B} = 0, \quad \nabla \times \mathbf{E} + \dot{\mathbf{B}} = 0,$$
$$\nabla \cdot \mathbf{D} = \rho, \quad \text{and} \quad \nabla \times \mathbf{H} - \dot{\mathbf{D}} = \mathbf{J}. \tag{3.1}$$

It is necessary to add to Maxwell's equations additional equations which relate the fields $\mathbf{E}$, $\mathbf{B}$, $\mathbf{D}$, $\mathbf{H}$. These are called constitutive relations (CR) because they model the constituents of the underlying matter. The most familiar CR are given by complex-valued, frequency dependent, permittivity $\varepsilon(\omega)$ and permeability $\mu(\omega)$ as

$$\mathbf{D}(\omega) = \varepsilon(\omega)\mathbf{E}(\omega) \quad \text{and} \quad \mathbf{B}(\omega) = \mu(\omega)\mathbf{H}(\omega) \tag{3.2}$$

These (CR) are suitable model for naturally occurring materials including water, glass and iron. However, the range of possible CR is much more extensive. Permittivity can be inhomogeneous, $\varepsilon(\omega, \mathbf{x})$, such as blocks of different media sandwiched together. More examples include anisotropic media, such as isospar, where the permittivity $\underline{\varepsilon}(\omega)$ is a tensor. This gives rise to birefringence. Chromium oxide can be shown to have a small magnetoelectric effect. Here, the displacement current $\mathbf{E}$ depends slightly on the magnetic field $\mathbf{H}$. We can write these as

$$\mathbf{D}(\omega) = \varepsilon(\omega)\mathbf{E}(\omega) + \alpha(\omega)\mathbf{H}(\omega) \quad \text{and} \quad \mathbf{B}(\omega) = \mu(\omega)\mathbf{H}(\omega) + \beta(\omega)\mathbf{E}(\omega) \tag{3.3}$$

The response of other media depends not only on the frequency ω but also the wavevector $\mathbf{k}$. Such media are known as spatially dispersive

$$\mathbf{D}(\omega, \mathbf{k}) = \varepsilon(\omega, \mathbf{k})\mathbf{E}(\mathbf{k}, \omega) \quad \text{and} \quad \mathbf{B}(\omega, \mathbf{k}) = \mu(\omega, \mathbf{k})\mathbf{H}(\omega, \mathbf{k}) \tag{3.4}$$

Such media are also known as non-local. This is because, after taking the inverse Fourier transformation of (3.4), one can observe that $\mathbf{D}$ as some point depends on the electric field $\mathbf{E}$ of other points. Another important case is non-linear CR. For non-linear permittivity, we set $\varepsilon(\mathbf{E})$ to be a function of $\mathbf{E}$. An example is non-linear crystal, which can lead to frequency doubling. An example of non-linear permeability $\mu(\mathbf{H})$ is for soft magnets, where the permeability falls from around $1000\mu_0$ to $1\mu_0$.

Modern manufacturing techniques allow us to tune CR for desired properties. Indeed, the range of possible CR is limited only by the ingenuity of scientists and engineers to build and test new materials and that of theoreticians to find mathematical methods to model them. Materials with CR that do not occur in nature are called artificial functional materials or metamaterials.

Ultimately, all CR are approximations of the complex underlying structure of the medium. One can imagine, for example, the very high electromagnetic fields that could exist inside individual atoms of a crystal. The complex interactions between the sub-atomic particles and incoming radiation are then averaged to produce an effective CR. As such, CR should be thought of as a tool for describing a medium. As a result, complex physics is replaced by a number of parameters. In the simplest case given in (3.2), where the medium is not magnetic $\mu = \mu_0$ and one works at a single frequency ω_0, the entire system is reduced to a single number $\varepsilon(\omega_0)$. As well as being simple, these approximations can be extremely accurate. Thus, CR should be thought of as a useful tool for describing the electromagnetic properties of a medium.

To be accurate, CR requires a large number of comparable substructures per wavelength of light. This homogenizes the electromagnetic fields so that CR can be used. Natural materials in the optical spectrum will have thousands of atoms per wavelength. Cells are used to create artificial functional material and metamaterials, and they must be repeated in a pattern. As a rule of thumb, for CR to be accurate, one needs at least ten cells per wavelength. When working in the microwave regime, the cells must be about a millimeter in size. Therefore, the structures within each cell must be smaller than a submillimetre in size. In addition, to avoid boundary effects, there should be at least ten wavelengths of material. Thus, a real three-dimensional material requires 100^3 cells.

The different CR described above (temporally dispersive, spatially dispersive, magnetoelectric, non-linear, inhomogeneous, etc.) may be considered as CR models. These models may also include models of the functions $\varepsilon(\omega)$, $\mu(\omega)$, etc., such as the Lorentz model and the Drude model. Having decided what CR model one should use, the next step is to find out the actual values. The more complected the model, the more parameters one needs to find. There are a number of ways of doing this. If one has a block of the material, then one may be able to measure the CR directly. If one assumes CR given by (3.2), one can look at the transmission and reflection coefficients, or the dispersion relation, at different frequencies to measure $\varepsilon(\omega)$ and $\mu(\omega)$. Another possibility, especially appropriate for artificial functional materials, is to put a single cell onto a computer and use an electromagnetic solver to calculate the dispersion relation and transmission and reflection coefficients. By imposing periodic boundary conditions is necessary only to put a single cell on to the computer.

If one has a model of the complex underlying structure, one can perform an averaging process on the electromagnetic fields $\mathbf{E}$, $\mathbf{B}$, $\mathbf{D}$ and $\mathbf{H}$ and derive the parameters that way. This process is called homogenization. The exact details of the homogenization procedure depend on the choice of the CR model.

Considering the simplest CR model given by (3.2), $\varepsilon(\omega)$ and $\mu(\omega)$ are the key parameters that define a material's response to an EM wave, we can categorize materials based on these parameters. Conventional materials with $\varepsilon > 0$ and $\mu > 0$, referred to as double positive media (DPM), include common dielectric materials such as polytetraflufoethylene, Al_2O_3, etc. Materials with one of the constitutive relations positive and the other are negative are single negative media (SNM), such as plasmas or metals with a negative permittivity (positive permeability). Unlike the double positive media, these single negative media only allow evanescent wave transport. The last case represents a special case of materials where both permittivity and permeability are simultaneously negative. These Double NeGative materials (DNG) like their double positive counterparts support wave propagation through the media. The key difference between the double negative quadrant and the other three is that single negative and double positive media all occur naturally, whereas we are yet to find a naturally occurring double negative media. Some authors have argued that Lobster eyes or moth eyes are DNG media, although as we discuss the size of the elements in these structures are comparable or larger than a wavelength in size and as such do not meet homogenization criteria necessary to truly act as a DNG media.

3.3.1 Artificial dielectrics

One of the earliest recorded propositions for an engineered artificial EM material was in the 1890s by Rayleigh, who proposed a system of small scatterers as an equivalent continuous medium [7], acting as an artificial dielectric. In 1914, Lindman considered small wire helices embedded into a host medium to create an artificial chiral material [8]. Although the first practical applications of artificial dielectrics did not appear until the 1940s with the pioneering work of Kock [9]. Kock created artificial dielectrics from arrays of sub-wavelength metallic structures (spheres, rods and plates) to form lightweight, compact, RF dielectric lenses [9].

In 1953, Brown [10] extended this work, examining a lattice of thin metallic wires (see inset of Figure 3.1), showing such a system could be considered to have a plasma frequency. Brown demonstrated that this wire array system forms an

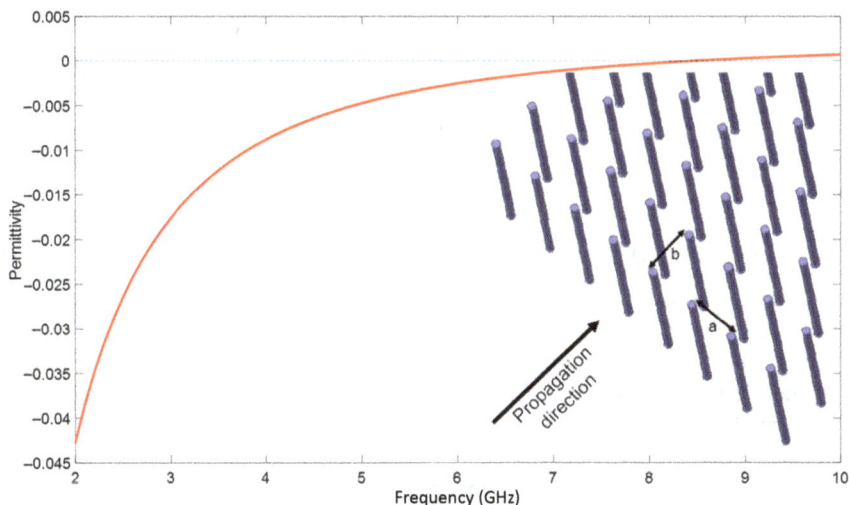

Figure 3.1 Permittivity of an effective wire array media, for $r = 40$ μm, $a = 10$ mm and $b = 5$ mm

effective medium with negative permittivity (positive permeability). Brown started by considering the case of lossless wires, meaning the wire array can be modelled as an array of inductors, with inductance L. This enabled Brown to model the permittivity for an artificial dielectric made from infinite long wires of radius r, separated by a distance b in the direction of propagation of normal incidence EM wave, and a distance a perpendicular to wave propagation is given by [10]

$$\varepsilon = \frac{\lambda_0}{2\pi b} \arccos \left[\cos\left(2\pi b/\lambda_0\right) + \frac{\lambda_0}{2a} \frac{\sin\left(2\pi b/\lambda_0\right)}{\ln(a/2\pi r)} \right] \frac{\tan(\pi b n/\lambda_0)}{Z_0 \tan(\pi b/\lambda_0)} \tag{3.5}$$

where λ_0 is the free-space wavelength of the incident wave, with the impedance of free space $Z_0 = 376.8\Omega$. Figure 3.1 shows a realization of the wire array media and the associated permittivity for the media; note below a certain frequency, the permittivity is negative, decreasing rapidly as the frequency decreases. Kharadly and Jackson [11] generalized this work to include effective media formed from lattices of metal ellipsoids, disks and rods, with the assumption that the frequency of operation is low and the Rayleigh quasi-static restriction holds.

Interest in effective wire media grew as the possibilities for exploitation were realized. Most comprehensively by Rotman [12] who explored these artificial materials as plasma analogs to investigate the effect of plasmas on antenna systems. This type of wire array media has also been turned into an 'active' material, by the inclusion of active components, enabling the media to be actively switched from a negative to a positive permittivity medium. Progress with this type of media resulted in the material becoming commercially available in the 1970s [13]. A material whose permeability is positive, and permittivity can be switched by the application of a current from positive (allowing wave propagation) to negative, acting as a plasma and reflecting incident EM waves. These materials were first proposed as antenna covers where the whole or specific sections of the cover could be made transparent for short periods of time. Even today, wire array media is still an active area of research, especially in configurations as materials that exhibit spatial dispersion, i.e. a dependence of the permittivity or permeability on the wavevector, $\varepsilon(\omega, k)$ and $\mu(\omega, k)$ [14–16], or as subwavelength elements for materials with simultaneous negative ε and μ, or as analogs for quantum mechanical barriers [17].

3.3.2   Artificial magnetic media

Following on from the successful research into artificial dielectrics, researchers started to consider artificial magnetic media. As in conventional materials, the magnetic field component of the EM wave couples only weakly to the material [18], there is a lot of interest in engineering materials that can couple strongly to an EM wave. The key sub-wavelength artificial magnetic meta-atom geometry is the split ring resonator (SRR) developed by Schelkunoff and Friis [19]. Although many researchers have investigated magnetic meta-atom [20,21] the base geometry remains the SRR, see Figure 3.2 where rings are formed from continuous metallic tracks formed into rings with at least one split gap in them. An EM wave incident

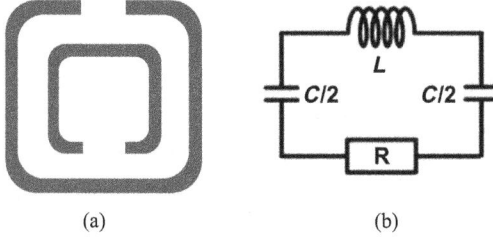

Figure 3.2   (a) Double SRR meta-atom geometry and (b) equivalent circuit for an SRR

on the SRR cause microscopic currents to flow in the metallic elements, producing a magnetic flux to oppose the incident field. Without the split, this interaction would be purely an inductive non-resonant phenomenon, resulting in a weakly diamagnetic system. The split prevents the current from circulating, causing a collection of charge at the split edge, creating a capacitance. A single SRR meta-atom will accumulate charge at the gap, creating a large electric dipole moment that, in most cases, dominates over the magnetic dipole moment. The addition of a second concentric SRR, where the 'gaps' of the SRRs are opposite each other, offers control over the capacitance of the meta-atom, allowing the electric dipole moment of the inner ring to suppress the electric dipole moment of the outer ring, allowing the magnetic moment to dominate the system.

Resonant frequency of the meta-atom $\omega_0 = \sqrt{1/(L + R/j\omega_0)C}$, allowing one to estimate the magnetic moment m_h of a meta-atom to first order as [22]. A double SRR, see Figure 3.2, to first order can be modelled as a quasistatic equivalent LCR circuit [23]. The inductive elements of the equivalent circuit are estimated by $L \approx 2\mu_0 r$, and the Ohmic loss as $R \approx \pi r/c\sigma\delta$. The capacitance consists of two components: a capacitance from the split 'gaps' and the capacitance between the individual split rings of an individual meta-atom, which can be approximated [24] by $C \approx \pi r \varepsilon_0 t/2d$, where t is the combined width of the rings and d the separation between the rings. Leading to an approximate

$$m_h(\omega) = \frac{\pi^2 r^4 \mu_0 H}{(\omega_0^2/\omega^2 - 1)L} \tag{3.6}$$

From which the effective permeability [22] (μ_{eff}) can be determined,

$$\mu_{\text{eff}}(\omega) = 1 + \frac{m_h}{VH} \tag{3.7}$$

where V is the unit cell volume of an individual meta-atom. This first-order approximation does allow useful insights into how engineered geometries define the effective permeability of the material.

3.3.3   Double negative media

Although several researchers considered materials with simultaneous negative constitutive parameters, permittivity and permeability, referred to as DNG materials [25], and materials with a negative index of refraction [26] prior to 1965. The first systematic study of the general properties of a hypothetical DNG medium with a negative refractive index is attributed to the seminal 1967 paper by Veselago [27]. These DNG-type materials, unlike materials with a single negative (SNG) constitutive parameters, support EM wave propagation, although unlike conventional media with both positive constitutive parameters (DPM) in a DNG material, the Poynting vector of a propagating wave would be antiparallel to the direction of the phase velocity, the 'energy' of the EM wave propagates in the opposite direction to k.

Veselago also predicted several other remarkable properties that would arise from a DNG medium. The reversed Doppler effect, where a detector in a DNG medium moving towards an EM source will measure a frequency lower than the frequency the wave was emitted with, not higher as would be the case in a DPS media. Then, there is the reversed Vavilov–Cerenkov effect, where a particle moving through a medium in a straight line, as shown in igure 3.3, will emit an EM wave, called Cherenkov radiation. The cone of this Cherenkov radiation cone is given by $\cos\theta = (\beta\sqrt{\varepsilon\mu})^{-1}$, where β is the normalized particle velocity. This means in the case of a DNG material, as the angle θ is obtuse (Figure 3.3), the Cherenkov radiation is 'backward' propagating.

The start of the 21st marked a turning point for interest in artificial materials with the publication of the seminal paper by Pendry [28]. Pendry's paper reconsidered the Veselago lens using transformative optics and presenting a mechanism that allows the diffraction limit to be beaten. Pendry pointed out how evanescent waves propagating in a DNG material can be redistributed in space such that the waves are transported far from the source[28]. The first realizations of a DNG material came from the work by Smith and Schultz, who constructed the first DNG media in 2000–2001. The basic form realized was a material with a negative refractive index in one direction of propagation [29]. This work was quickly followed by the famous two-dimensional NIM [30]. Where each subwavelength unit

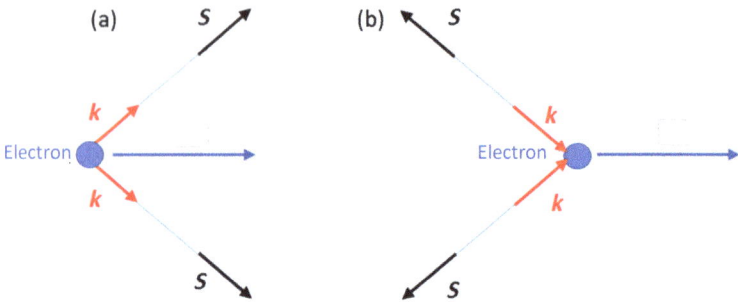

Figure 3.3   (a) Cerenkov effect in a DPM, (b) Cerenkov effect in a DNG material, where S is the Poynting vector and k is the wave vector

Figure 3.4   (Left) Conventional double SRR on FR4, middle image, double CSRR cut into Cu, (right) image array of metamaterial formed from double SRR on Kevlar with wire array on the back

cell consists of two basic elements, a double SRR supported on a dielectric substrate (FR4) with a Cu track (wire) placed behind the substrate. Each unit cell had an overall size of 3 mm; the unit cells were fabricated and arranged into an array of 3 by 20 by 20 unit cells, forming the DNG material.

This approach of combining an SRR to give the negative permeability response, with a conducting wire strip acting as a shunt inductance, remains the most common approach to creating DNG materials. Although this configuration does require a way to support the two elements' electrical succinct from each other. Usually achieved by placing a dielectric substrate between the wire strip line and SRR. An alternative approach developed by [31] is to derive a planar negative of the SRR in metal; this complementary SRR (CSRR) is shown in Figure 3.4. Where the gaps cut into the metal are the complement of the SRR tracks. In a CSRR, the capacitance's of the SRR are replaced by inductors, resulting in the CSSR presenting a negative permittivity response to an incident EM wave, exhibiting a behaviour which is almost dual to that of the SRR. More recently, a range of other alternative geometries have been used to construct DNG unit cells, such as electromagnetically induced transparency geometries [32].

3.4   Applications

The possible applications for metamaterials are endless and cannot easily fit into a single chapter of a book on general electromagnetics. In this section, we discuss some of the key technological applications metamaterials have been applied to and considered for. However, do remember this list is not exhaustive.

3.4.1   Antennas

Ultimately, antennas are the interface between propagating EM waves and electronic circuits. Enabling the transmission of data, where excited charge carriers in the antenna radiate EM waves, or the reception of data, by the excitation of charge carriers in the antenna driven by the interaction with an incident EM wave creating current flow. The EM wavelengths a specific antenna can interact with are dependent on the antennas geometry, for example, in general, an antenna that is small compared to the incident EM wavelength will reflect most of the incident wave. A metamaterial antenna behaves as if it were much larger than its actual size, because its novel structure stores and re-radiates energy.

3.4.2   Lenses

The use of metamaterials and metasurfaces for lenses has two main goals: one is to create perfect lenses to produce images with resolution beyond the diffraction limit of light, and the other is to create thinner lenses.

Perfect lenses were first proposed by Veselago in the 1960s [27] and significantly promoted by Pendry in the 2000s [28]. Pendry observed that one could exploit ideas from general relativity and model the double-negative lens in terms of an optical metric. Here, spacetime has been folded on itself. He demonstrated that not only do the propagating modes correctly focus, but so do the evanescent modes; thus, the double-negative metamaterial has anti-evanescent modes. The uses of such lenses would be extensive. One could probe the evanescent fields at some distance from the source, locating sub-wavelength structures. Thus, producing the same advantages as near-field scanning optical microscopy, at an arbitrary distance from the sample.

There are multiple challenges to creating a perfect lens. Not only do the permittivity and permeability have negative, but these must be equal. Such negative constitutive relations are usually achieved near resonances and tend to be very narrow band. This leads to the other problem. All materials have losses, however near resonances, these materials are very lossy. Unlike the evanescent modes, which are reversed, the losses due to the materials cannot be eliminated. It is ongoing research to see if active materials can reverse these material losses without introducing noise [33,34].

The most successful demonstration of perfect lenses is by using silver or gold. A 120-nm-thick silver lens that was placed 60 nm below a patterned mask [35]. Using light of 700nm, details half that size could be observed. However, both the source and target need to be within a wavelength of the light, so this does not achieve the goal of arbitrary distance.

Metasurfaces [36] as an alternative to zone plates also known as Fresnel diffraction lenses. Both are very thin and have advantages and disadvantages [37]. Metasurface lenses can be tunable [38,39], remove chromatic distortion [40] and broadband [41].

3.4.3   Active materials

A new possibility opens up when one considers material where the permittivity changes in time. In this case, we can write $\varepsilon = \varepsilon(t)$. Such a material can be made using an array of diodes or a non-Forster network [42, 43]. The most exciting possibility here is considering the case when one has a bulk medium where the permittivity is controlled and is changed for the entire medium from positive to negative at one moment. A simple analysis of the situation states that the waves pass from being propagating waves $\mathbf{E} = \mathrm{Re}(\mathbf{E}_+^I e^{-i\omega_I t + ikz} + \mathbf{E}_-^I e^{-i\omega_I t - ikz})$ to exponentially growing and shrinking $\mathbf{E} = \mathrm{Re}(\mathbf{E}_+^{II} e^{-\omega_{II} t + ikz} + \mathbf{E}_-^{II} e^{\omega_{II} t - ikz})$. Here, the scripts (I and II) refer to before and after the change in permittivity. Temporal boundary conditions $[\mathbf{B}] = 0$ and $[\mathbf{D}] = 0$ can be used to link the coefficients $\mathbf{E}_+^I$ and $\mathbf{E}_+^{II}$. We note that such a response does not break the conservation of energy as the system requires an external DC current in order to keep the permittivity negative. The ability to use a DC current to exponentially grow an arbitrary signal would have vast potential, and hence, there is much active research to be the first to demonstrate this phenomenon.

Although this analysis works for media without damping such a model of the media is not physical. Care must be taken when dealing with a media with damping, especially if such a media is modelled by a constant permittivity, such as a loss tangent. This model is not consistent with a time-dependent medium as it would predict non-physical response. Repeating the above when the passing from vacuum to a positive permittivity with a loss tangent actually predicts exponential growth, although in this circumstance, there is no external source of energy [44].

Another suggested use for temporal media is as a spacetime cloak. This is the relativistic version of the stealth material described below in section 3.4.5. Such a medium can then be used to 'hide time' as well as space [45,46]. What this means in practice is that one can hide an object for a brief moment in time. In [47], the authors demonstrated experimentally that they can hide a few picoseconds, enough for a signal to pass unnoticed by an observer. The theoretical work was extended [48] to include dispersive media.

3.4.4 Electromagnetic wave profile shaping

The concept of spatial dispersion can be used to help design waveguides to enhance the interaction between electromagnetic waves and particle beams. A simple calculation [49] shows that a three-dimensional waveguide can be modelled by a one-dimensional spatially dispersive medium. Likewise, a wire array can also be modelled by a wire medium can also be modelled by a one-dimensional spatially dispersive medium [51]. By varying the dimensions of the waveguide or the radii of the wires, the medium becomes inhomogeneous. Such media forms a useful test bed for investigating media that are both spatially dispersive and inhomogeneous. The usual constitutive relations for studying spatial dispersive media, $\varepsilon(\omega, \mathbf{k})$ (3.4), are incompatible with the constitutive relations required for inhomogeneous media $\varepsilon(\omega, \mathbf{x})$, since the variables $\mathbf{x}$ and $\mathbf{k}$ are Fourier conjugate variables of each other. In [15, 16, 50], it is pointed out that one method of prescribing the constitutive relations is in terms of differential equations relating the longitudinal polarization $P(\omega, z)$ and the longitudinal electric field $E(\omega, z)$. We start with the undamped hydrodynamic Lorentz model, which is spatially dispersive and homogenous. The constitutive relation [14, 51, 52] for this model in the frequency–wavevector representation is usually written as

$$P(\omega, k) = -\frac{\omega_P^2 \, \varepsilon_0}{\omega^2 - \omega_0^2 - \beta^2 c^2 k_z^2} E(\omega, k) \tag{3.8}$$

where $c\beta$ is the polariton velocity, ω_0 is the polariton resonance frequency and ω_P is the plasma frequency. Taking the inverse Fourier transform with respect to k_z, we obtain the differential equation

$$c^2\beta^2\frac{\partial^2 P}{\partial z^2} - (\omega^2 - \omega_0^2)P(\omega,z) = -\varepsilon_0\omega_P^2 E(\omega,z) \tag{3.9}$$

Working in physical, as opposed to wavevector, space, enables us to allow the parameters ω_P and ω_0 to become functions of position. Setting $D = \varepsilon_0 E + P = 0$, we obtain solutions to electromagnetic wave propagation. These satisfy the ODE

$$c^2\beta^2\frac{\partial^2 P}{\partial z^2} - (\omega^2 - \omega_0(z)^2 - \omega_P(z)^2)P(\omega,z) = 0 \tag{3.10}$$

The quantity $\omega_0(z)^2 + \omega_P(z)^2$ corresponds to the plasma cut-off frequency, which is a function of the radii of the rod or the dimensions of the waveguide. As such, these profiles can be chosen, and thus, one can tailor the solutions to Maxwell's equations and, thus, the longitudinal shape of the electromagnetic waves.

3.4.5   Stealth materials

No application of metamaterials has captured the imagination of researchers, industry, defence and the public more than the concept of creating cloaking devices. The principle behind creating a 'cloak' is to reduce the total scattering cross section of the object to be cloaked, essentially creating an area of no, or zero, EM field, which requires a DNG material to achieve [53]. Limited space prevents a full treatment of cloaking in this chapter, but we would refer the interested reader to the report by Alitalo and Tretyakov [54], which in addition to a good overview of cloaking also discusses applications.

The first form of cloaking discussed is based on the transformation of the spatial coordinates to create a region of space with zero field [55]. Transformation optics is a design approach, as opposed to a material structure; it allows the user to define a suitable distribution of the permittivity and permeability in the media [56] to give a specific, desired form of EM wave propagation through a region of space filled with the defined material. After defining the required permittivity and permeability distribution using the techniques discussed previously, the user would design a meta-atom to yield the required permittivity and permeability. Transformation optics is an approach to solving EM wave propagation in inhomogeneous materials utilizing the invariance of Maxwell's equations under coordinate transformation, provided the constitutive relations are transformed appropriately. Meaning transformation optics can be used to find a distribution of the permittivity and permeability for a medium that encompasses a region of space to 'curve' an EM wave around the region, essentially cloaking an object in that region. The difficulty with this approach is that the required permittivity and permeability distribution needs a material where $0 < \mu < 1$, $0 < \varepsilon < 1$ and $\mu = \varepsilon$, which requires the use of resonant sub-wavelength geometries, such as the SRR to achieve. This means although transformation optics is a very powerfully approach to designing a metamaterial cloak, inherently, the media forming the cloak has a very narrow bandwidth of operation, making fabrication very complex and difficult [56].

An alternative to field-transformation cloaking is the use of transmission-line media [57] to form a cloak. Created from a network of either a simple mesh of transmission lines (unloaded) or from a mesh of capacitors/inductors forming a reactive loaded network [57]. Where both types of networks are designed with regions of space that have effectively zero field [57]. Where cloaking arises by restricting wave propagation to the parts of the volume that are filled by the transmission-line network. Media formed from unloaded transmission lines, where the permittivity is constant over the range of frequency's cloaking is required, are essentially free of dispersion, creating the possibility of a broad-band cloaking material. Although the phase delay in these media is a factor of $\sqrt{2}$ larger than that of free space, creating an EM wave phase mismatch between the cloaking media and free space, resulting in the cloak casting a shadow. In loaded transmission-line media, the reactive elements of material can be designed such that the material phase constant matches that of free space. This removes the shadow effect seen in unloaded media cloaks but introduces frequency dispersion of the effective parameters, drastically reducing the possible bandwidth of operation.

There are obviously a number of applications for a cloak that can "hide" a complete volume; here, we focus on two applications, first an application for a simplistic fabricated transmission line network cloak, and second, an application of low-observable materials to reduce EM reflection to hinder detection and lock-on technologies.

The first case we present examines the use of a transmission line network to reduce scattering from support structures blocking antenna apertures.

Space satellites and spacecrafts generally have a very limited surface area for mounting antennas, creating situations where antennas can be blocked, for example, by metallic support structures.

In this case, the proposed transmission-line cloak can be used to mitigate the blockage of these structures, simplifying the task of designing the spacecraft. The types of struts/blocking objects that can be cloaked with this method are naturally limited by the fact that the object must fit inside the TL cloak, as discussed before. Also, with the known method of coupling waves from free space to the cloak (TL network) [19], the cloaking can be achieved only for one polarization. This means that the antenna that is blocked must emit only linearly polarized radiation.

References

[1]   Mossotti F. *Memorie di Matematica e di Fisica della Societa Italiana delle Scienze*, 1850.
[2]   Clausius R. Die *Mechanische Warmtheorie*. 1879.

[3]   Maxwell Garnett JC. "Colours in metal glasses and in metallic films," *Phil Trans. R Soc Lond.* 1904;203:385–420.

[4]   Bruggeman DAG. "The prediction of the thermal conductivity of heterogeneous mixtures," *Ann Phys.* 1935;24:636–664.

[5]   Lakhtakia A. (Ed). *Selected Papers on Linear Optical Composite Materials*, Bellingham, WA: SPIE Press; 1996.

[6]   Belov P and Simovski C. "Homogenization of electromagnetic crystals formed by uniaxial resonant scatterers," *Phys Rev E.* 2005;73:1–9.

[7]   Rayleigh. "On the influence of obstacles arranged in rectangular order upon the properties of a medium," *Phil Mag.* 1892;5(34):481.

[8]   Lindman F. *Ofversigt af Finska Vetenskaps-Societetens Förhandlingar.* 1914;*LVII*, A(3).

[9]   Kock WE. "Metallic Delay Lenses," *Bell Syst Tech J.* 1948;27:58.

[10]   Brown J. *J Proc IEEE.* 1953;100:625.

[11]   Kharadly MMZ, and Jackson W. "The properties of artificial dielectrics comprising arrays of conducting elements," *Proc IEE.* 1962;100:199–211.

[12]   Rotman W. "Plasma simulation by artificial dielectrics and parallel-plate media," *IRE Trans Ant Prop.* 1962;10:82–95.

[13]   Chekroun C, Herrick D, Michel YM, Pauchard R, and Vidal P. "Radant–new method of electric scanning,". *L'Onde Electrique.* 1979;59(89).

[14]   Belov P, Marques R, Maslovski SI, *et al.* "Strong spatial dispersion in wire media in the very large wavelength limit," *Phys Rev B.* 2003;67:113103.

[15]   Boyd T, Gratus J, Kinsler P, *et al.* "Customizing longitudinal electric field profiles using spatial dispersion in dielectric wire arrays," *Opt Express.* 2018; 26:2478–2494.

[16]   Gratus J and McCormack M. "Spatially dispersive inhomogeneous electromagnetic media with periodic structure," *J Opt.* 2015;17(2) 025105.

[17]   Riga J and Seviour R. "Electromagnetic analogs of quantum mechanical tunneling," *J Appl Phys.* 2022;132:200901.

[18]   Landau L Liftshitz E and Pitaevskii. *Electrodynamics of Continuous Media.* Oxford: Pergamon; 1984.

[19]   Schelkunoff SA and Friis HT. *Antennas: Theory and Practice.* Oxford: Wiley; 1952.

[20]   Hardy WN and Whitehead LA. "Split-ring resonator for use in magnetic resonance from 200–2000 MHz." *Rev Sci Instrum.* 1981;52:213–216.

[21]   Schneider HJ and Dullenkopf. "Slotted tube resonator: A new NMR probe head at high observing frequencies," *P. Rev Sci Instrum.* 1977;48:68.

[22]   Marques R FM and Rafii-El-Idrissi R. "Role of bianisotropy in negative permeability and left-handed metamaterials," *Phys Rev B.* 2002;65:144440.

[23]   Pendry JB, Holden AJ, Robbins DJ, *et al.* "Magnetism from conductors and enhanced nonlinear phenomena," *IEEE Trans Microw Theory Tech.* 1999; 47:2075–2084.

[24]   Baena JD, Marques F, Martel J, *et al.* "Artificial magnetic metamaterial design by using spiral resonators" *Phys Rev B.* 2004;69:014402.

[25]   Siukhin DV. *Opt Spektrosc.* 1957;3:308.

[26] Schuster A. *An Introduction to the Theory of Optics.* London: Edward Arnold; 1904.

[27] Veselago VG. "Electrodynamics of substances with simultaneously negative electrical and magnetic permeabilities," *Sov Phys Usp.* 1968;10(4):504–509.

[28] Pendry JB. "Negative refraction makes a perfect lens," *Phys Rev Lett.* 2000; 85:3966.

[29] Smith DR, Willie J. Padilla, Vier D, *et al.* "Composite medium with simultaneously negative permeability and permittivity," *Phys Rev Lett.* 2000; 84:4184–4187.

[30] Shelby RA, Smith DR, and Schultz S. "Experimental verification of a negative index of refraction," *Science.* 2001;292:77–79.

[31] Falcone F, Lopetegi T, Baena JD, *et al.* "Effective negative-e stopband microstrip lines based on complementary split ring resonators," *IEEE Microw Wirel Compon Lett.* 2004;14:280.

[32] Liao Z, Liu S, Feng H, Li C, Jin B, and Jun Cui T. "Electromagnetically induced transparency metamaterial based on spoof localized surface plasmons at terahertz frequencies," *Sci Rep.* 2016; 6:27596.

[33] Xiao S, Drachev VP, Kildishev AV, *et al.* "Loss-free and active optical negative-index metamaterials," *Nature.* 2010;466(7307):735–738.

[34] Khurgin JB. "How to deal with the loss in plasmonics and metamaterials," *Nat Nanotechnol.* 2015;10(1):2–6.

[35] Melville DO, Blaikie RJ, and Wolf CR. "Submicron imaging with a planar silver lens," *Appl Phys Lett.* 2004;84(22):4403–4405.

[36] Zou X, Zheng G, Yuan Q, *et al.* "Imaging based on metalenses," *PhotoniX.* 2020;1:1–24.

[37] Engelberg J, and Levy U. "The advantages of metalenses over diffractive lenses," *Nat Commun.* 2020;11(1):1991.

[38] Roy T, Zhang S, Jung IW, *et al.* "Dynamic metasurface lens based on MEMS technology," *APL Photonics.* 2018;3(2) 021302.

[39] Arbabi E, Arbabi A, Kamali SM, *et al.* "MEMS-tunable dielectric metasurface lens," *Nat Commun.* 2018;9(1):812.

[40] Khorasaninejad M, Aieta F, Kanhaiya P, *et al.* "Achromatic metasurface lens at telecommunication wavelengths," *Nano Lett.* 2015;15(8):5358–5362.

[41] Wan X, Xiang Jiang W, Feng Ma H, *et al.* "A broadband transformation-optics metasurface lens," *Appl Phys Lett.* 2014;104(15) 151601.

[42] Pacheco-Pena V, Kiasat Y, Solls DM, *et al.* "Holding and amplifying electromagnetic waves with temporal non-Foster metastructures," arXiv:230403861.2023.

[43] Engheta N. "Four-dimensional optics using time-varying metamaterials," *Science.* 2023;379(6638):1190–1191.

[44] Gratus J, Seviour R, Kinsler P, *et al.* "Temporal boundaries in electromagnetic materials," *New J Phys.* 2021;23(8):083032.

[45] McCall MW, Favaro A, Kinsler P, *et al.* "A spacetime cloak, or a history editor," *J Opt.* 2010;13(2):024003.

[46]   Kinsler P, Gratus J, McCall MW, *et al.* "Dispersion in space-time transformation optics," In: *2016 URSI International Symposium on Electromagnetic Theory (EMTS)*. Piscataway, NJ: IEEE; 2016. pp. 356–358.
[47]   Fridman M, Farsi A, Okawachi Y, *et al.* "Demonstration of temporal cloaking," *Nature*. 2012;481(7379):62–65.
[48]   Gratus J, Kinsler P, McCall MW, *et al.* "On spacetime transformation optics: temporal and spatial dispersion," *New J Phys*. 2016;18(12):123010.
[49]   Siaber SS, Gratus J, Seviour R, *et al.* "Corrugated waveguide with matched phase and group velocities: an extended regime of wave-beam interaction," *Opt Express*. 2024;32(13):23288–23302.
[50]   Gratus J, Kinsler P, Letizia R, *et al.* "Electromagnetic mode profile shaping in waveguides," *Appl Phys A*. 2017;123:1–6.
[51]   Song W, Yang Z, Sheng XQ, *et al.* "Accurate modeling of high order spatial dispersion of wire medium," *Opt Express*. 2013;21(24):29836–29846.
[52]   Ciraci C, Pendry JB, and Smith DR. "Hydrodynamic model for plasmonics: a macroscopic approach to a microscopic problem," *ChemPhysChem*. 2013;14(6):1109–1116.
[53]   Alu A and Engheta N. "Plasmonic materials in transparency and cloaking problems: Mechanism, robustness, and physical insights," *Opt Express*. 2007;15(6):3318–3332.
[54]   Alitalo P and Tretyakov S. "Metamaterials for space applications," ESA, Ariadna ID: 07/7001d; 2008.
[55]   Pendry JB, Schurig D, and Smith DR. "Controlling electromagnetic fields," *Science*. 2006;312:1780–1782.
[56]   Kildishev A and Shalaev V. "Transformation optics and metamaterials," *PhysUsp*. 2011;54(1): 53–63.
[57]   Alitalo P, Luukkonen O, Jylha L, *et al.* "Transmission-line networks cloaking objects from electromagnetic fields," *IEEE Trans Antennas Propagat*. 2008;56(2):416–424.

Chapter 4

Future developments in superconducting motors and flux pumps

Tim Coombs[1]

The history of electric motors goes all the way back to 1821, when Faraday demonstrated that a wire, dipped in mercury on which a magnet was placed, would move in a rotating motion when a current was passed through it. By 1827, Jedlik had demonstrated a device that required no permanent magnets and incorporated a stator, a rotor and a commutator and by 1855 had developed the technology far enough for it to be incorporated into a model electric vehicle.

At its most basic level, an electric motor requires current and a magnetic field. The interaction between these produces a force. This force can either be used to produce a torque and hence rotation or directly to provide linear motion.

A key parameter is power *P*.

For rotating motion, the governing equation is

$$P \propto BIV\omega$$

where B is the magnetic loading, I is the armature (current) loading, V is the volume and ω is the rotation speed.

For the applications discussed here, power density is more important than the power. Thus, in the absence of the discovery of new lighter materials, we can eliminate volume from our discussion.

Additionally, there is a limited return from increasing the speed as, unless we are going to use a gearbox (which itself would add weight and reduce reliability), the speed tends to be dictated by the application.

Consequently, if we wish to make a step change and produce significantly higher power density motors, the primary options are to increase the magnetic loading and/or the armature loading. Superconductors are the ideal candidates for both options.

Superconductors have been around for more than a century. They are principally used in applications where they provide a clear advantage over conventional materials or where there is simply no alternative. Thus, they are used in magnetic resonance imaging (MRI) and nuclear magnetic resonance (NMR) devices,

[1]Department of Engineering, University of Cambridge, UK.

research magnets in laboratories or particle accelerators such as the large hadron collider. All of these applications are principally down to the ability of super-conductors to provide a high magnetic flux far in excess of that available from permanent magnets, continuously and in a relatively compact space.

In recent years, there have been intensive efforts to use some of the other properties of superconductors to develop high-power density motors and generators, bearings for energy storage flywheels, superconducting electronics, fault current limiters, transmission lines, etc.

Superconductors are well known for their high current densities and ability to transfer electricity without loss (under DC conditions).

Thus, we have a material that can provide a magnetic field larger than that available from permanent magnets, support a current orders of magnitude higher than conventional conductors and (having no loss) enable a high-efficiency machine.

One of the major drivers towards the effort to produce superconducting motors is that there is a strong focus within the research and development organizations of the commercial aviation sector on the improvement of aircraft emissions and noise. ACARE Flightpath 2050 [1], equivalent to NASA N+3 goals, set reduction targets of 75% CO_2, 90% NO_x and 65% perceived noise emissions compared to the year 2000 baseline. Although a significant portion of these reductions are expected to be met by improvements in engine technology, it is likely that a switch to electric propulsion will be required actually to achieve these targets. There have been various schemes proposed but most of these envisage a system of distributed electric propulsion usually supported by a single gas turbine.

Assuredly with their unique properties superconducting motors are the ideal vehicle for what would be a revolution in air travel.

Two problems must be solved before this can be brought about. The first is that a superconductor is only lossless in DC conditions, so if they are to be used in a fully superconducting motor then consequential AC losses must be minimised.

The second is more subtle. Since a motor depends on the interaction between magnetic fields and current, if the motor is to be made as power dense as possible the stator current needs to be maximised and the rotor field also needs to be maximised.

Unless the rotor is acting as a permanent magnet, as would be the case if it were constructed from superconducting mini-magnets [2–10], we would require sliding contacts, e.g. solid metal-graphite blocks, metal-plated carbon fibers, metal fibers or liquid metal systems [11]. Although these exist, they are relatively unreliable and are best eliminated if possible.

Currently, to eliminate the AC loss problem, most of the research has concentrated on synchronous machines in which only the rotor is superconducting and therefore only sees a DC field and current. However, this limits the gains. Only half the machine is superconducting, and consequently, additional efforts are being made to make conductors where the loss is minimised.

We need a step change in the motors and specifically the design of the conductors that are being used to make them before they can be made all superconducting. This chapter summarises the progress so far in superconducting motors, where they are going next, and introduces the idea of flux pumping as one way to eliminate the need for sliding contacts.

4.1 History

The first recorded instance of a superconducting motor was in Newcastle by the International Research & Development Co (a subsidiary of Parsons) led by Tony Appleton. The project, which started in 1963, was a homopolar motor which is based on the Faraday Disc principle. It was a 50 hp (37 kW) motor running at 2000 rpm. It was commissioned in 1966 and was a prototype for a 3000 hp motor running at 200 rpm, which was seen as the breakeven point for commercial viability. The ultimate application was seen to be ship propulsion motors. The motor was tested in 1970/1971 and ran successfully. The main weakness of the homopolar design is the need for sliding contacts (carbon brushes at the time), which introduce large losses and require a lot of maintenance. Although homopolar machines were not abandoned altogether, and as recently as 2000, there was a US research project for a 300 kW ship propulsion motor, and more information can be found in [12], this line of development has been overtaken by other designs.

The next large motor of note is the AMSC synchronous motor, which was also developed for ship propulsion. There were two stages to this program a 5-MW prototype followed by a 36.5 MW full-scale motor. Weighing 75 tonnes, the 36.5 MW version used BSCCO (a high-temperature superconductor [HTS]) rather than NbTi, and its mass compares to 280 tonnes for the equivalent induction machine and 400 tonnes for a synchronous machine. The machine was delivered for testing in 2007. It passed all the tests. Figure 4.1 taken from [13] illustrates the powerful case for an HTS motor in terms of reduced size and weight.

The shift to synchronous motors is significant. A homopolar motor is a DC machine and, as far as the rotor is concerned, a synchronous motor incorporates a DC element. There are two aspects to advanced motors. The first is size and weight and a superconducting motor with potentially much increased armature loading and field strength enables significant weight/size reductions, as can be seen from the graphic. The second is efficiency. If you apply an AC magnetic field or an AC current to a superconductor, then there is a loss. Since this loss occurs at a low temperature, thermodynamics tells us that the loss needs to be multiplied by a factor that is dependent on the temperature: the lower the temperature, the higher the factor.

A synchronous machine in which the superconductor is on the rotor mitigates the loss since, for the most part, the superconductor sees a DC current and field.

This means that the motor can be reduced in size and weight as was seen with the AMSC motor. However, the machine is only 'half superconducting' and there are considerable advantages to be gained if the machine can be made wholly superconducting. Thus, the next generation of motors is set to be wholly superconducting.

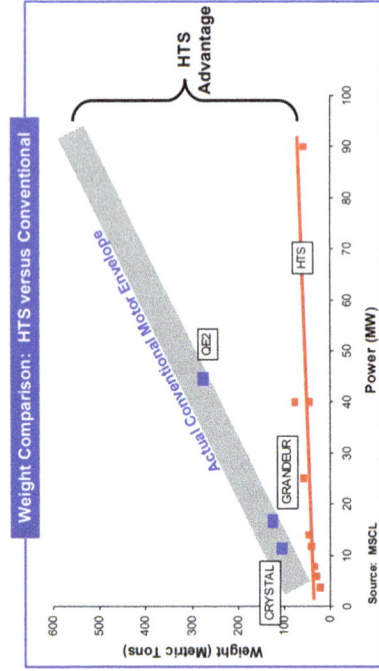

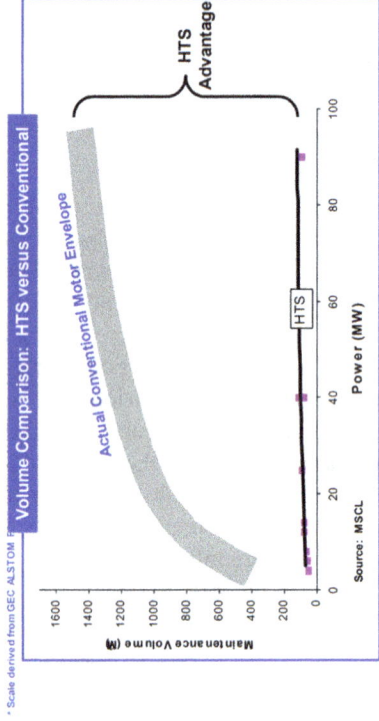

Figure 4.1   Volume and weight comparisons for HTS versus conventional

Fundamentally, a wholly superconducting machine is only really practical if the AC loss problem can be solved or at a minimum mitigated. Since the AC loss is primarily a magnetisation loss, the route to the reduction of this loss is either cables in which the magnetic field is cancelled, such as in Roebel cable, or one in which the filament width is minimised. Minimising the filament width works because the loss is a function of flux, phi (not B), which is the product of B and the area; hence, thinner twisted filaments equal lower loss. An alternative method that has been explored is to use a magnetic shield to protect the superconductors from seeing a varying magnetic field.

Minimising filament widths in conductors like Nb_3Sn, BSCCO or MgB_2 has been well researched. NbTi development in particular was well developed, when, as early as 1973, in [14] Gallagher-Daggitt gives an excellent review of what came to be known as Rutherford Cables.

To make the motor as small as possible, we need the highest current, and this in itself presents a problem as the highest critical currents are found in the (RE) BCO family, which are tapes (and hence by definition do not have narrow widths). Many potential solutions have emerged (e.g. Roebel tapes, CORC, striation, etc.)

As well as traditional wound rotors and stators, superconducting bulks can be used as permanent magnets.

Although the original motivation behind research into superconducting motors was for ship propulsion, the focus has switched to aircraft propulsion. The challenges are considerable and success will depend on a power-to-weight ratio of greater than 20 kW/kg. This is a target that simply cannot be met without superconductors.

One other motivation behind the move towards electric aircraft is the possibility of using liquid hydrogen to replace jet fuel. Superficially, this further enables superconducting technology since the fuel will also require a cryogenic environment.

Thus, the future of superconducting motors may well be powering electric aircraft, but there are a few problems to solve before that is realised.

4.1.1 Materials

4.1.1.1 Available materials

The generally accepted price for superconductors to compete on price with copper is 10\$/kAm. Given in Table 4.1 is a guide to the relative prices of a range of superconductors. The table was presented by Mike Tomsic of Hypertech in a presentation entitled 'Commercial Rules for Making Money with Superconducting Applications' at the NIST/DOE Workshop on Enabling Technologies for Next Generation Electric Machines in 2015 [15]. He presented two tables first was for an iron-cored machine in which the field is limited to 2 T and the second reproduced here (Table 4.1) for an air-cored machine running at 6 T (we are assuming that air is more important since the weight will be the primary driver).

It can be seen that both Nb_3Sn and NbTi had reached the price threshold and that the price is projected to be stable. None of the so-called HTS materials was yet

Table 4.1   Price performance at 6 T and 10 T for an air core machine.

Price-performance $/kAm for motors and generators	Present $/kAm	Projected future 3 years $/kAm	Projected Future 6 years $/kAm
MgB2 4K-6T	50	3.5-10	3.5
MgB2 10K-6T	70	4.9-15	4.9
YBCO 20K-6T	80	40	16
YBCO 30K-6T	120	60	25
BSCCO 2223 4K-6T	125	125	125
Nb3Sn 4K-6T-ITER type	4.90	4.90	4.90
Nb3Sn 8K-6T-ITER type	9.80	9.80	9.80
Nb3Sn 4K-6T-ITER type	11.70	11.70	11.70
Nb3Sn 4K-6T high Jc tube type	2.17	2.17	2.17
Nb3Sn 4K-10T - high Jc tube type	4.68	4.68	4.68
Nb3Sn 8K-6T - high Jc tube type	4.34	4.34	4.34
Nb3Sn 4K-6T - optimized APC tube type			0.60
Nb3Sn 8K-6T - optimized APC tube type			1.20
Nb3Sn 4K-10T - optimized APC tube type			0.75
NbTi 4K-6T	1.00	1.00	1.00
NbTi 4K-10T	10	10	10

cheap enough although the price was projected to fall. The only exception is BSSCO which is neither cheap enough nor expected to fall: this is because the price of BSSCO is dominated by the price of silver, a vital component in its manufacture.

Hence, on cost alone, the choice is obvious and favours low T_c Nb_3Sn or NbTi, but that is not the whole story, and in any event, the price is expected to drop.

The above assumes that the material is used in cable or wire form. Another form that is available is bulk. These are lumps of HTS in which the material is used as if it were a permanent magnet. The lumps can be grown (e.g. melt processed YBCO) or fabricated from sections of tape cut up and stacked on top of each other, but in either case, they can act as very compact mini-magnets. Very high fields in excess of 17 T (peak) have been achieved in a 3cm diameter cylinder.

In principle at least, a superconducting PMM could provide a very compact motor. However, there are two barriers to this: the first is that the PMMs need to be magnetised in situ. This could be done with the field windings (perhaps), but the other problem is more intractable and that is the PMMs tend to demagnetise, principally due to the cross-field effect. Nevertheless, there are research efforts in progress to try to solve these problems.

4.1.2   Motivation

4.1.2.1   Efficiency

Electric motors are very efficient: typically $> 95\%$ for a large motor. In 2017, ABB issued a press release claiming an efficiency of 99.05% [16] for a conventional synchronous motor. Since this is a 44 mW motor, the release claims a saving of $500,000 over its 20-year life. A similar superconducting motor could achieve an even higher efficiency, and indeed, it has been shown that if electric-powered long-distance flight is to be achieved, then the efficiency needs to be greater than 99.95% [17].

4.1.2.2   Weight – aircraft

It has long been an aspiration to produce the more-electric aircraft and to follow that with the all-electric aircraft. To achieve that there are considerable technical barriers to be overcome. Superconductors present an opportunity to increase the power-to-weight ratio.

The power of an electric motor may be expressed in terms of its electrical loading, its magnetic loading, its volume and its rotation speed. With conventional electrical materials, both electrical loading and magnetic loading are limited. Unless a gearbox is used, then the rotation speed is limited by the thrust required from the motor. In addition, mechanical constraints limit the rotation speed, and these become more stringent as the motor gets bigger. Larger motors develop more power due to their increased volume, but they will also be heavier. All of these constraints lead to maximum power-to-weight ratios of less than 20 kW/kg (Table 4.2).

Thus, in principle at least, superconductors that can sustain higher magnetic fields and higher electrical loading are the only way to dramatically increase the power-to-weight ratio. Much has been made of superconducting mini-magnets that have been magnetised up to 17 T [7,10]. However, there are constraints even here. The first is that there is no practical way to generate 17 T in situ in a motor. The second is that the cross-field effect has a deleterious effect on the mini-magnets' magnetisation and the final one is that, practically, the motor will require a magnetic circuit that limits the usable flux density to the order of 2 T.

We are therefore left with electrical loading, which in principle gives superconductors a huge advantage, but here there is also a problem: AC losses. A fully superconducting machine will have superconductors subjected to a varying current and magnetic field. Both of these exposures will cause losses. This reduces the efficiency and ultimately the magnitudes of both current and field which can be used.

4.1.2.3   Volume – ships

Electric propulsion has many potential advantages. A comprehensive list can be found here [19]. This source lists many potential advantages, not least the reduction in vibration and noise which is achieved by separating the prime mover and propeller shaft, but there is another potential advantage that is common to electric machines in general and superconducting ones in particular and that is the potential reduction in volume.

While weight is important in aircraft, it is less so in ships. However, there are strong arguments that can be deployed as to why a reduction in volume would be highly beneficial. The first argument, but largely a specious one, is that it provides

Table 4.2 Large brushless motor parameters from various manufacturers

	Maximum power (kW)	Maximum torque (N m)	Maximum speed (RPM)	Maximum voltage (V)	Peak efficiency (%)	Maximum weight (kg)
magniX	640	3,020	2,300	800	—	200
Safran	500	—	2,200	—	—	—
EMRAX	380	1,000	6,500	800	Up to 98	42
H3X	250	120	20,000	800	95.7	15
Siemens	204	1,500	1,300	—	—	49
YASA	200	790	8,000	700	>95	37
MGM COMPRO	80	300	12,000	800	~95	22
Pipistrel	57.6	220	2500	400	—	22.7
Plettenberg	50	142	5,600	320	>93	12.5
Thin Gap	44	191	2,200	726	—	4
T-Motor	28	56	3,134	100	—	5.2
MAD Components	21	67	—	100	—	4.2
ePropelled	16	30	8,000	72	94	3.1
KDE Direct	12.8	—	—	69.6	—	0.7
Turnigy	9.8	—	7,800	52	—	2.5
Alva indus-tries	10	37	4,000	—	95.4	2.8

Source: Reference [18].

0° Shaft Angle

10° Shaft Angle

Shaft Angle	Performance Loss
3°	0.14 %
5°	0.39 %
10°	1.52 %
15°	3.41 %

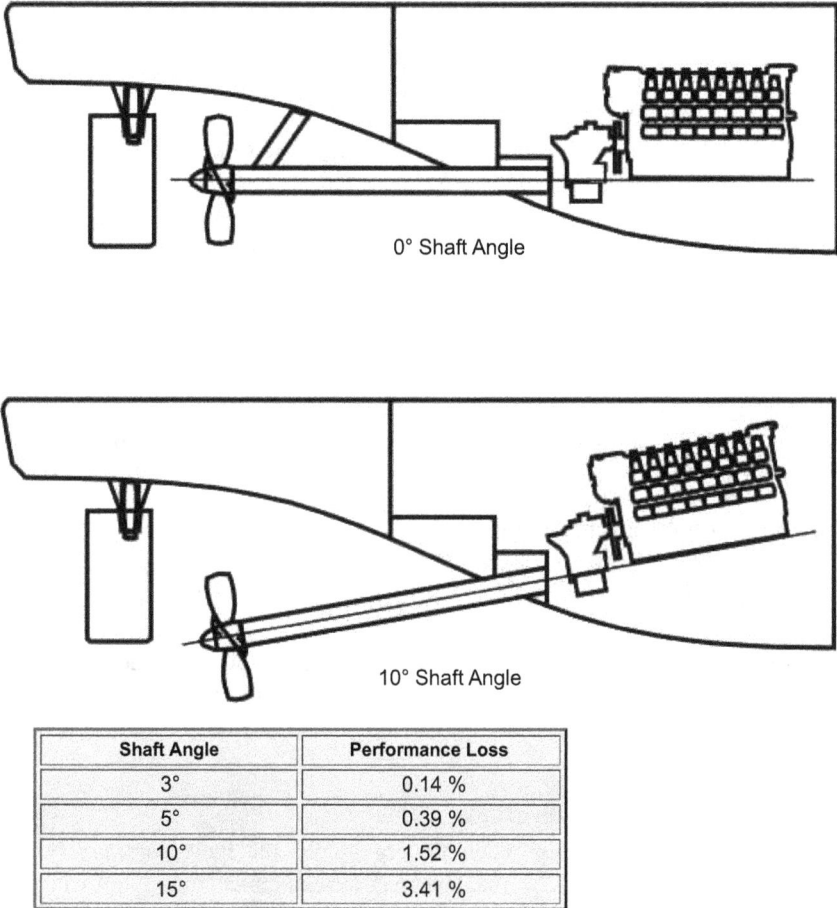

Figure 4.2   Effect of shaft angle on efficiency. Source: PROPULSION SYSTEMS (ricepropulsion.com) [20].

more space for the payload. While this may be true, of far greater importance is the position and angle of the drive shaft. For an inboard motor, the larger the motor, the greater the angle of the drive shaft from the horizontal. As Figure 4.2 shows, this leads to a performance loss from the propeller.

If podded propulsion (housing of the motor in a streamlined pod external to the hull) is used, then there is still a strong incentive to reduce the diameter of the motor as this will reduce the drag from the pods.

4.1.3   Types

The big problem with superconductors is the presence of AC losses. Any losses which are incurred at low temperatures are magnified by thermodynamics. To understand this, we need to refer to the following simple equation for the Carnot efficiency:

$$\eta = \left(\frac{T_{cold}}{T_{hot} - T_{cold}}\right) \times 100\%$$

Thus, if we are operating at, for example, 4.2 K and the room temperature is 293 K. The maximum Carnot efficiency is 1.45%. However, this is the ideal efficiency in practice a perfect Carnot engine cannot be made. There are several cooling systems available. For example, Gifford-McMahon, Brayton, pulse tube, Stirling cycle, etc., but a typical efficiency might be ~30% [21]. Taking this example means that removing 1 W of heat occurring at 4.2 K would require 230 W of power.

Hence, it is clear that we need to avoid/minimise losses at low temperatures. This has led to early developments in superconducting motors, concentrating on those in which the superconductor sees only a DC field. There are two main types: homopolar and synchronous.

4.1.3.1   Homopolar machines

Homopolar machines are extremely simple; they rely on the fact that a current passing through a magnetic field will experience a force that is perpendicular to both (Figure 4.3).

The basic construction is simple; the major drawback is the need for sliding contacts to transfer the current into and out of the machine, but the superconductor which is providing the B-field sees no AC input; hence, research continues on this type of machine.

4.1.3.2   Synchronous machines

Superconducting synchronous machines are by far the most common type of superconducting machine. The machine has a superconducting rotor which, in principle at least, sees a constant magnetic field. There are two possibilities for the

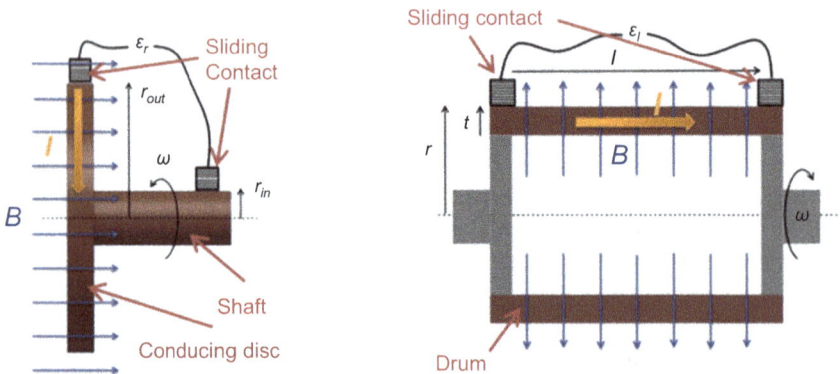

Figure 4.3   Basic homopolar motor geometries [22]

rotor. The first is a permanent magnet-type rotor, and the second is a powered rotor. The first can be realised using superconducting mini-magnets, which are either bulks (single crystal lumps of superconductors) or stacked tapes, which are "bulks" constructed from multiple layers of superconducting tape. Both types are capable of sustaining extremely high magnetic fields: the record is >17 T [7,10] for both types. Also attractive is that the bulks are typically 3 cm across, so they can be tiled onto a round rotor [23].

There are three drawbacks to bulks. The first is the process of magnetisation. The world record was achieved by placing a bulk inside a magnet with a field slightly greater than the record field. The superconductor was then cooled, and the external field ramped down. These conditions cannot be practically replicated inside a motor. The most common way of magnetising the rotor is to use the field coils to provide a pulse field. Under these conditions, the maximum usable flux density is of the order of 3 T [48].

The second drawback is that the quoted 17 T is the peak flux density, not the average flux density. The field is shaped like a circular cone. The magnetic pressure is a function of B^2. Hence, the force from a bulk is one-half of the force that would be available if the average field were 17 T.

Cylindrical field for radius R and peak field B_{max}

$$F_{cylinder} = \frac{B^2}{2\mu_0} \pi R^2$$

The conical field for base radius R and peak field B_{max}

$$B = B_{max} \frac{r}{R}$$

$$F = \int_0^R \left(B_{max} \frac{r}{R}\right)^2 \frac{2\pi r}{2\mu_0} dr$$

$$F_{cylinder} = F_{cone}/2 = \frac{B^2}{2\mu_0} \pi \frac{R^2}{2}$$

A permanent magnet would be limited to the order of 1 T, but its field is approximately uniform. Hence, a bulk field and a permanent magnet field produce equivalent forces once the bulk field exceeds approximately SQRT(2) T, which is less than the 3 T quoted by Durrell.

The third problem is the so-called cross-field effect. Even in a synchronous machine, the field the rotor sees is not constant and fluctuates. This will cause an AC loss, but more seriously, if there is a component perpendicular to the axis of magnetisation (the so-called cross field), then this will gradually demagnetise the rotor [24].

The alternative is to have a wound rotor. In a study for a 36 MW ship motor (PowerPoint Presentation (cern.ch): https://indico.cern.ch/event/763185/contributions/3415632/attachments/1897527/3143412/Wed-Mo-Po3.12_-_SC_Synchronous_Motors_for_Electric_Ship_Propulsion.pdf) the Table 4.3 data are presented.

Table 4.3   Comparison of SC machines for ship propulsion.

	LTS field coil Assembly	AMSC HTS Field coils[1]	Homopolar inductor Alternator, HTS[2]
Output power (MW)	36.5	36.5	36
Speed (r/min)	120	120	120
Number of poles	30	16	18
Terminal voltage (kV)	6	6	3.8
Armature current (A)	3600	1270	
Efficiency (%)	99	97	>95
Mass (ton)	43	75	100
Length (m)	1.8	3.4	4
Outside diameter (m)	4.3	4.1	2.9
Armature cooling	Forced air	Liquid	

[1]B. Gamble *et al.* Full power test of a 36.5 MW HTS propulsion motor, *IEEE Trans. Appl. Supercond.*, (2011).
[2]K. Sivasubramaniam *et al.* High power density HTS iron core machines for marine applications, *IEEE PES*, (2007).

The quoted mass for the LTS motor is 43 tonnes; hence, the power density is relatively low at 0.84 kW/kg, not useful for aircraft propulsion, but the 99% efficiency is attractive for ship propulsion. One could envisage (with LTS) the rotor coils operating in persistent mode and therefore not requiring power, eliminating the problem inherent in homopolar machines of the need for sliding contacts. This would not necessarily be ideal though, as the rotor field would be fixed.

4.1.3.3   Induction/synchronous machines

In 2010, Nakamura *et al.* [25] presented a motor with a BSCCO squirrel cage winding that operated in both induction and synchronous mode. Essentially, the rotor was magnetised during the induction phase as the motor was driven up to speed. Once it was up to speed, the currents in the rotor persisted (with small losses due to joint resistance), and the motor operated as a synchronous machine. They claim a tenfold increase in torque over the conventional rotor that they replaced. They built a 20 kW machine, and there are a series of papers that trace its development. The most recent of which is [26].

This is a good interim step and potentially at least removes the problem of powering the rotor. However, to reach truly high power densities, the focus needs to shift to fully superconducting machines.

4.1.3.4   Fully superconducting machines – induction motors

To date, there are very few fully superconducting motors [23]. This is perhaps not surprising given the problems with AC losses. There have been a few outline designs, and there is an ARPA-E project which started in 2022 to develop one. There are however many attempts at producing low-loss windings in order to mitigate the problem and thereby facilitate a fully superconducting machine.

To understand this, we need to understand the sources of the loss. There are many references that describe the various sources of loss and the different loss regimes, but the main mechanisms are hysteresis due to magnetisation and demagnetisation and coupling losses either between filaments (if present) or to the normal conductors (if present) in the wire. In extremis, there can be a flux flow loss when a significant number of the filaments are driven into the flux flow state by the coupling currents.

CORC, CROCO and Roebel are candidates for HTS cables. More simply, for conductors that can be constructed from filaments such as BSCCO and MgB_2, there are solutions that depend on making the superconducting filaments as fine as possible. There are also efforts to striate HTS tapes.

All of these are candidates, but to date they do not present a solution. They are either wasteful of tape and difficult to fabricate (Roebel) or wasteful of space, as is the case with CORC, where the tape is wound onto a non-conducting tube. Striation has potential and will cut down on AC losses by changing the geometry of the flat tape into filamentary conductors such as can/has been done with BSCCO and MgB_2. Coupling between the strands (since they cannot be twisted) is not eliminated, reducing the effectiveness.

Detailed analyses of the reasons for utilising multi-filamentary conductors are given in [27–29].

A possible solution is to produce a hybrid Roebel tape where, instead of cutting patterns in tapes and then threading them together, the tapes are themselves patterned, and then multiple layers are created by depositing a second and a third layer of superconductor, each with a different pattern (Figure 4.4).

Multiple layers of REBCO can be deposited, as reported by Ha *et al.* [31]. Thus, we can use this patterning technique to create a facsimile of a Roebel tape from a single tape. This represents the most efficient manufacturing method as the successive layers can be deposited sequentially on a single production line, and there is no cutting or assembly as is required for the other cable fabrication methods reported above.

4.2   Summary of Developments

To date, there have been many efforts to develop superconducting motors. These efforts are driven by the need for smaller, lighter, more efficient machines. However, if they are ever to gain traction in the market, three things need to happen.

The first is cost; superconductors are expensive and currently are unable to deliver on cost. There are targets for the price: typically, these would hover around the 10 dollars per kiloampmetre level. It is true that the cost has been reduced, if this measure is used, but this is a misleading figure. The price per metre of conductor tends to stay the same, reducing only gradually. The big advances, for coated conductors anyway, have been in the current carrying ability.

A chain is only as strong as its weakest link, and this is true of coated conductors. Along the length of the superconductor, the critical current will vary, and

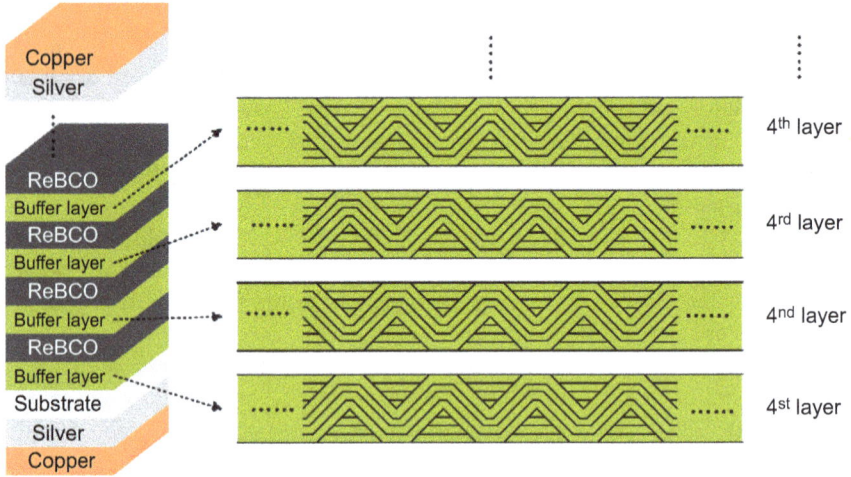

Figure 4.4   Patterned superconductors for AC loss minimisation and current maximisation (PSALM) [30]

the selling price is dictated by the minimum current along its length. True, this could be ameliorated by making long lengths from multiple sections, but this leads to the need for costly jointing procedures and losses in the joints. Better processing has led to higher currents over longer lengths.

The second is the problem of AC losses, which limit the practicality of all superconducting machines. A synchronous machine, which is only half super-conducting, is still limited by the capabilities of conventional materials and, if machines that have power densities of 40 kW/kg or more are to be made, they will have to be all superconducting.

The final problem is the provision of the current itself. If we are to fully utilise the capabilities of high-current superconductors, then we need to be able to find a way of feeding those currents into the machine. Current leads are one solution, but high current leads are bulky and this limits the possibility of reducing the size of the machine. In addition, high-current power supplies are heavy, bulky and expensive.

A possible solution is flux pumping and this will be discussed in the next section.

4.2.1   Flux pumping

Flux pumps are a means of providing current to a coil or a magnet without direct electrical or physical contact. Typically, a MRI coil, for example, would be charged directly using current leads, which would then be withdrawn if the magnets oper-ated in the persistent mode (as with magnetic resonance imaging) or left in place if the field required maintenance or ramping. Flux pumps are not appropriate for an MRI constructed from LTS in which the current leads are generally removed once the magnet is charged. LTS magnets have low to no joint resistance, and they are

DC devices so there are minimal losses and there is no need for direct power. The purpose of a flux pump is to provide power during operation. An MRI constructed from HTS however would benefit from a flux pump to maintain the field as HTS joints do not have zero resistance, and in a large coil, there will be many joints. Efforts are underway to construct HTS MRIs incorporating flux pumps [32], and it is estimated that these will be ready for production in the next 3 years. Table 4.4 provides a summary of the flux pump technologies.

4.2.1.1 MRI magnets

Before describing flux pumps themselves, it is worth noting the impact that flux pumps could have on MRI magnets. These advantages will also be accrued when they are used in motors, but HTS MRIs are much closer to market, do not have the same problems with AC losses that are found in motors and themselves represent an evolutionary step on the road to fully superconducting HTS motors.

Flux-pumped ultra-high current magnets have the potential to enable the efficient use of the extremely high currents that REBCO tapes are capable of sustaining. This in turn enables the construction of ultra-compact magnets. A perennial use of superconductors is an enabling element in MRI magnets. However, traditional MRI is bulky, heavy and typically requires the patient to be brought to the MRI rather than the other way around.

Currently, CT (computer-aided tomography) is the mainstay for acute imaging in traumatic brain injury (TBI), ischaemic stroke [33,34] and intracranial haemorrhage. However, it does not possess the soft tissue contrast provided by MRI, which can detect stroke within minutes to hours of onset, and identify important injury mechanisms in TBI. Such diagnoses can be invaluable for early prognostication as well as for stratification of patients for clinical trials of novel drugs and conventional therapies. Initial experience with dedicated emergency MRI departments is encouraging [35].

Mobile near-patient MRI for acute brain imaging is a tantalizing prospect, but no clinically usable prototype solution is currently available. The technological challenges are substantial. The magnetic field in traditional MRIs is generated by low-temperature superconductors which operate in liquid helium (@4.2K). Complicated heat insulation requirements and the weight of magnets mean that conventional MRIs typically weigh in excess of several tons. Making a portable MRI with conventional technology is not considered practical. High-temperature superconductors operating at higher temperatures have much higher current carrying capability, making lighter magnets possible. However, currently, HTS magnets have to be driven by bulky power supplies via thick warm-to-cold current leads, which complicates heat insulation, limits current capacity, and affects field stability. Flux pumping an HTS magnet provides an approach to eliminate the reliance on such power supplies and current leads, opening up the way to a truly compact, transportable and low-cost MRI [36], which could be taken to the patients' bedsides. Further weight savings can be achieved by developing an ultra-short magnet. Although this would inevitably reduce the volume of the uniform field region, having a magnet that would be able to traverse the brain axially and be

Table 4.4 Summary of flux pumping technologies

Method	Characteristics	Pros	Cons	Notes
Moving normal spot	Flux is drawn into a superconducting circuit via a moving normal spot	Possibly the simplest method to understand. A normal spot is induced in the superconductor which contains fluxons. As the spot moves across the superconductor the fluxons move with it	Of limited use. Both the source and the spot must move together	This method requires only a single spot
Thermally induced switches	Flux is permitted to enter a superconducting circuit by the removal of a section of the superconducting circuit (typically by heating)	The spot is widened to include the whole width of a section of superconductor. Flux can then pass through this section. A second switch allows the flux to move to the load. The switches are operated sequentially	Superconductor must be heated to its transition point. This limits the frequency of operation of the device	Low-frequency operation leads to decay between cycles and hence a higher ripple field. 50 Hz operation has been achieved but only at relatively low currents
HTS travelling wave	Flux flow in the superconductor creates areas of different impedance	Superconductivity is never destroyed. In principle at least. Removing the requirement for heating	The system is not lossless. There will be a loss associated with the motion of the flux and the development of the charging voltage. In addition, the current in the load cannot be separated from the current through the switch. Thus, the switches impedance is not independent of the voltage developed	

(Continues)

| HTS rectifier type | Flux traverses into a superconducting circuit when the superconductor is full of current and hence unable to resist the incursion | In this system, the current through the superconducting switch can be controlled independently of the load. There is no requirement for heating since superconductivity is never destroyed. Hence, the system can operate at a high frequency | The system is not lossless. There will be a loss associated with the motion of the flux and the development of the charging voltage | |
| Power electronic switches | In essence, this is not a flux pump although the operation is similar. It is simply a half or full-wave rectifier | MOSFETs (or similar) are readily available for high currents and have high switching speed | MOSFETS in common with all power electronics have on-state resistance. | Ripple is large due to on-state resistance of MOSFETs |

rotated around the axis of the brain would enable 3D scanning. This configuration would also reduce the number of gradient coils required.

High-temperature superconducting (HTS) ReBCO-coated tapes are now capable of producing superconducting currents at current densities and strains sufficient for fabricating high-field magnets in a compact space. Further, the state of the art in coil winding technology has progressed such that the latest HTS flux pump technology can enable compact and inexpensive high-field magnets, making them available to universities and research organizations worldwide for high-field MRI instruments and devices.

Finally, since the length of the conductor required to produce a particular field is inversely proportional to the operating current, the conductor is a substantial part of the cost of a magnet (currently, a typical price is £70 per metre), so less conductor equals a cheaper magnet.

4.2.1.2   Motors

In applications such as motors where you have to contend with not only joint resistance but AC losses on top, there are strong arguments for using a flux pump.

Flux pumps present two distinct advantages. The first is that there is no direct connection to the external world and, therefore, no associated heat loss. The second is that, although high current can be developed in the magnet or the coil, these can be produced without the need for high-current power supplies, which are bulky and expensive.

Flux pumps are especially appropriate for machines constructed from high-temperature superconductors (HTSs). They facilitate the full use of the HTS's current carrying capability, cheaply and simply, enabling smaller, lighter, and more powerful magnets, motors, and generators. In addition, and possibly uniquely, they can make use of the geometry of coated conductors in that the large surface area provides an ideal vehicle for the application of a flux pump.

4.2.1.3   Normal spot

At its simplest level, a flux pump operates in the following manner. The whole circuit is superconducting and is divided into two sections, left and right. A normal spot is created in the left-hand section, and flux is introduced. Then, a normal spot is created between the left- and right-hand sides, and flux is introduced to the right-hand side. The process is repeated multiple times with further flux being added with each cycle. Figure 4.5, which is based on Figure 4.1 of [37], illustrates the process.

4.2.1.4   Switch controlled

A short note was written by Atherton and published in *Cryogenics* in 1967 [38], which illustrates the potential of flux pumping. In this short note, he points out that as long as the switches are opened completely (rather than simply using a moving normal spot) and that, consequently, there is no current through the switches, then the theoretical efficiency is 100%. In practice, this limit is not practically reachable as the normal way of operating the switches is to heat them rather than physically

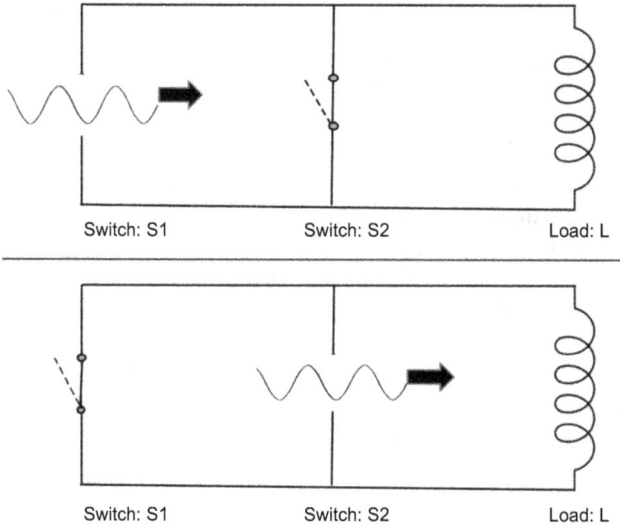

Figure 4.5   Flux pumping

open them, the resistance of the switch is therefore finite. Nevertheless, by reducing the current through the switches before they are opened, the overall efficiency can be made very high.

Although Atherton's note and Ten Kate's review imply that superconductivity has to be interrupted in order to enable flux pumping, this is not in fact the whole story.

There are two main classes of flux pump that do not require the interruption of superconductivity. One is the so-called travelling wave flux pump, and the other is the class of which the members depend on the dynamic voltage.

4.2.1.5   Travelling wave

The first of these is especially appropriate for motors since it uses a moving magnetic field, which could be present in the rotor of a motor (due to its interaction with the stator). Thus, the problem referred to above with getting power to the rotor is potentially solved. This effect, which was discovered by Coombs, has led to a number of efforts to integrate this mechanism into a motor or generator. The most active group is the Robinson group, who have dubbed their latest devices superconducting dynamos and, in a paper [39] delivered in August 2020, presented a study on its effectiveness in a 737-type aircraft taking a c. 500 km flight between Auckland and Wellington. Their conclusion was that a superconducting dynamo would enable a reduction in fuel burn of 0.47% in comparison to that which would occur if a conventional DC power supply were used.

4.2.1.6 Dynamic voltage

The dynamic voltage was first observed by Risse *et al.* [40]. This was reported as an ohmic DC resistivity. In fact, it is a voltage that occurs due to the motion of flux, and the phenomenon is fully explained in [41]. In their paper, Mikitik and Brandt point out that if flux is applied to a slab that is carrying a transport current, the consequent asymmetry in the current distribution means that the flux will transit across the slab, leaving from the opposite side to the one into which it entered. This change in flux produces a voltage, which is DC. This "dynamic voltage" is often referred to as dynamic resistance as it requires the presence of the transport current, and this is how it was first reported in [40], but this is misleading, as it is indeed a voltage and that voltage can be used to charge a coil.

Figures 4.6 and 4.7 illustrate the principle. They show a Bean model [42] representation of a slab or sheet. In Figure 4.6, there is no transport current. When a field is applied, currents are generated in the slab, which acts to screen the

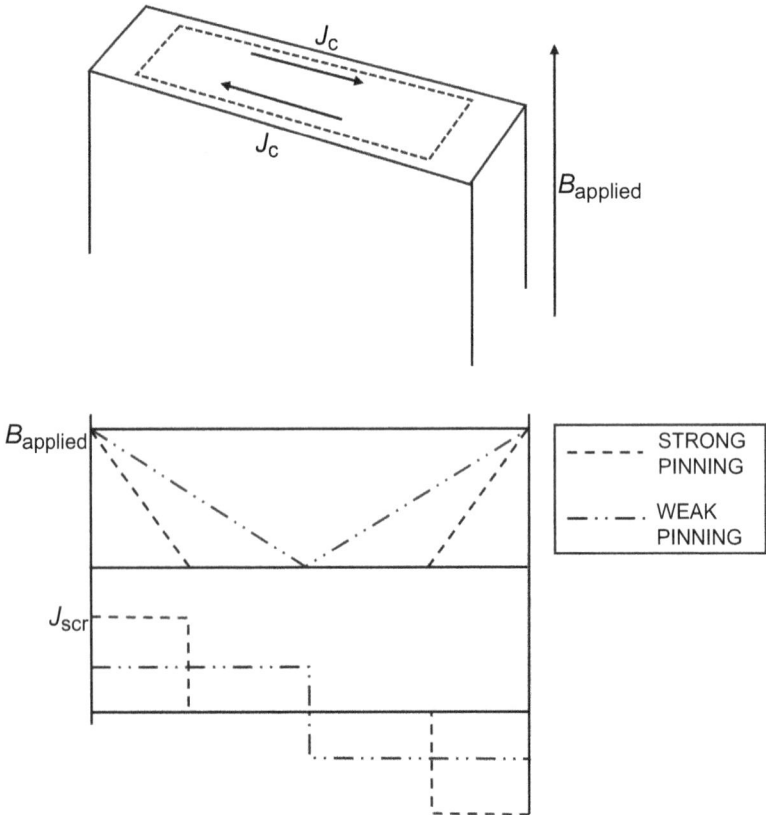

Figure 4.6   Bean model with no transport current

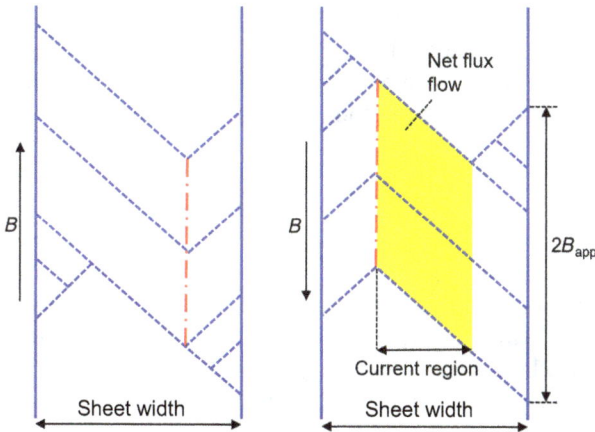

Figure 4.7   Bean model with transfer current – lateral motion of flux

interior of the superconductor by cancelling the applied field. The interior field distribution is symmetrical, and when the field is removed, the direction of the currents reverses. Note that the superconductor is being magnetised and demagnetised by the exterior field, but there will still be a hysteretic loss associated with the process.

The same thing happens if there is a transport current applied (Figure 4.7), but now symmetry will be lost. When the field is removed, the electric centre, as represented by the red chain-dashed line, moves across, so that flux moves laterally, and a voltage is generated.

4.2.2   Significance

High-temperature superconductors operating at higher temperatures than their low-temperature equivalents have much higher current carrying capability than either copper or low-temperature superconductors. This means that the coils used in motors or magnets can be made much lighter and smaller. This current carrying capability is very high indeed such that providing it would require bulky power supplies connected to our motor via thick warm-to-cold current leads. In turn, this complicates heat insulation, limits current capacity and affects field stability and in short negates at least part of the advantage provided by the use of the superconductors in the first place. Additionally, HTS tapes are currently very expensive and failure to exploit their full current carrying capability increases the overall cost of the system.

Flux pumping an HTS coil provides an approach to eliminate the reliance on such power supplies and current leads and enables the use of increased currents and thereby more power.

It may be some years before fully superconducting motors are available for use in aircraft, ships or other forms of transport but efforts are underway to dramatically reduce the size of MRI machines such that they can be made effectively portable, and it may well be that it is this research which paves the way for superconducting motors themselves. Another driver for superconducting technology is the move towards compact HTS tokamaks. These are at an early stage and it may be many years before they become practical but currently they are driving the price of HTS down.

4.2.3   Future

The superconductivity market is dominated by Asia Pacific Countries (APAC) and North America. Recent analysis by Technavio [43] shows that the market for Second-Generation HTS wire has surpassed that of LTS wire. The price of LTS is not dropping and the market for it, which is dominated by MRI and NMR, is largely mature. Factors, which are affecting the price of 2G wire, include methods of production, volumes of production and the opening up of new markets as it supplants LTS. "New" markets such as transmission and generation and Data Centres [43] as the demand for power increases worldwide may well be the drivers in the future.

Even though the cost problem may well be solved in the near future (<5 years), there remains the perennial problem of AC losses, which currently restrict the use of superconductors to DC applications.

As has been demonstrated, in applications, such as wind turbines, superconducting machines could be competitive if the cost is reduced. Trying to make a higher power density rotating machine requires the use of superconductors on both the stator and the rotor. This runs into the problem of AC losses, meaning that the machine may not have an efficiency high enough for an application such as aeroplanes. Even nuclear fusion, the current driver for HTS, would require a low loss AC cable if we are not going to run into the problem of catastrophic failure of our magnet through quench.

Notwithstanding the inherent problems, there have been some interesting developments in the recent past that relate to improvements in the tape such as PSALM and the building of new magnets and the prevention of quench.

For instance, a collaboration between the Lawrence Berkeley National Laboratory and the National High Magnetic Field Laboratory in 2021 [44] yielded the world's first HTS BSCCO ($Bi_2Sr_2CaCu_2Ox$) accelerator dipole magnet. In this context, HTS enable more powerful magnets than LTS (such as Nb–Ti) due to the higher upper critical field (BC_2 above 100 T at 4.2 K). The final magnet was based on a canted cosine-theta design using Bi-2212 Rutherford cables that were made from 0.8 mm wires, with a dipole field of 1.64 T in a 31 mm bore at 4.2 K. In addition, the magnet has excellent operating properties, with a quench current of 3600 A and a field hysteresis below 0.1%. This serves as a promising initial benchmark for HTS dipole magnets.

In addition, Molodyk *et al.* [45] present a new YBCO HTS tape for tokamaks with Y_2O_3 nanoparticles that can be easily produced over a long distance. This novel tape has excellent engineering current density (JE), with reported values of 1000 A/m^2 at 20 K/20 T and over 2000 A/m^2 at 4.2 K/20 T with a 40-μm substrate. Their new tape formulation involves adding Y_2O_3 nanoparticles that are native to YBCO as the vortex pinning centres, which contrasts with the commonly adopted method of introducing extrinsic nano-columns aligned about the c-axis as pinning centres. Consequently, the simplified approach resulted in over 300 km of YBCO wire manufactured in 9 months for a commercial fusion power customer. This simplicity is a key step in achieving the economies of scale needed to lower the price of HTS wires to an acceptable price for fusion and widespread adoption.

Moreover, an interesting project currently underway is the EU-funded SuperEMFL [46] which aims to produce 32+ and 40+ T all-superconducting magnets. Magnets at this scale are typically hybrid, comprising a superconducting coil with an internal copper coil. A pitfall of this hybrid construction is the excessive power consumption (several tens of megawatts) and the limited time of magnet operation. The SuperEMFL explores the use of an HTS insert coil, initially developed in the LNCMI-CNRS laboratory in Grenoble and discussed in [44], which set the record for the field generated by an HTS insert with a bore diameter

Table 4.5 The way ahead for HTS

	Reduced cost	**Reduced AC loss**	**Market forces**
Rotating machines	The high HTS cost is not offset by the benefits such as weight savings and simpler turbine construction	AC losses are a crucial problem in superconducting machines. Unless AC losses are reduced to manageable levels applications such as motors and generators will be problematic	Climate change is a key market driver for the wider adoption of HTS machines as they offer emissions-free operation despite the high cost
Nuclear fusion	The current costs of superconductors are not a key hindrance. Most fusion projects are funded by governments. Breakthroughs in fusion prototypes will lead to inevitable cost reduction of HTS due to economies of scale	Mitigating the AC loss problem improves the energy (and cost) efficiency of fusion magnets as they can operate closer to their critical current	Interest in fusion is highly motivated by the possibility of having clean, long-term energy source, driven by the need for addressing climate change and finding alternatives to fossil fuels
HTS MRI	The high cost of HTS is a key barrier to the widespread adoption of HTS MRI coils	AC losses are not a significant problem as coils are DC-operated	Pressing market drivers include the need for light, compact, and portable MRI. With HTS magnets these targets can be achieved. Consistent increases in helium prices affect LTS MRI systems

Significance: Highest ▇ Medium ▇ Lowest ▇.

of 38 mm. The insert was able to operate in a central field of 32.5 T, of which 14.5 T was generated by the insert itself. Cuprates (HTS) were used for the insert as they remain superconducting in much higher background fields than conventional LTS such as Nb–Ti. The insert used metal-as-insulation windings [47] to ensure stable operation and reduce the risk of irreversible damage from local HTS failure.

4.3   Conclusion

As to the way ahead, we present Table 4.5, which summarises some possible different applications and seeks to present which would be enabled if (1) the cost is reduced, (2) the AC loss problem is solved and (3) nothing happens but market drivers become the dominant force. Included in the table is fusion since, while fusion is still a long way off, the characteristics of HTS mean that HTS is a strong candidate for incorporation into a fusion reactor, and this could become the dominant application for HTS.

In conclusion, in this chapter, we have summarised the main applications of HTS superconductivity and the key enablers which will need to be found or made for the field to move forward. The two main factors are cost, which acts as a barrier to the entry; and AC losses, which act as a block on certain utilizations. A great deal of effort is being spent on both these factors, and as [47] reveals in Table 6 of his paper the magnet costs for LTS reactors rose from 37% of the total reactor cost in 2018 to 45.9% in 2020 so the goal may still be receding for LTS, even as we are learning. It remains to be seen whether the same is true for HTS reactors.

References

[1]   Flightpath 2050 Europe's Vision for Aviation, European Commission, Publications Office of the European Union, 2011.
[2]   G. Fuchs, G. Krabbes, P. Schätzle, *et al.*, *Appl. Phys. Lett.* 76, 2107 (2000).
[3]   M. Tomita, and M. Murakami, *Nature* 421, 517 (2003).
[4]   N. H. Babu, Y. Shi, K. Iida, and D. A. Cardwell, *Nat. Mater.* 4, 476 (2005).
[5]   A. Yamamoto, H. Yumoto, J. Shimoyama, K. Kishio, A. Ishihara, and M. Tomita, *23th International Symposium on Superconductivity*, Tokyo, 2010.
[6]   T. Naito, T. Sasaki, and H. Fujishiro, *Supercond. Sci. Technol.* 25, 095012 (2012).
[7]   J. H. Durrell, C. E. J. Dancer, A. Dennis, *et al.*, *Supercond. Sci. Technol.* 25, 112002 (2012).
[8]   G. Fuchs, W. Haessler, K. Nenkov, *et al.*, *Supercond. Sci. Technol.* 26, 122002 (2013).
[9]   J. Durrell, A. Dennis, J. Jaroszynski, *et al.*, *Supercond. Sci. Technol.* 27, 082001 (2014).
[10] A. Patel, A. Baskys, T. B. Mitchell-Williams, *et al.*, *Supercond. Sci. Technol.* 31, 09LT01 (2018).

[11]   R. E. Witkowski, F. G. Arcella and A. R. Keeton, *IEEE Trans. Power Appar. Syst.* 95(4), 1493–1500 (1976), doi:10.1109/T-PAS.1976.32246, accessed 29/12/24.

[12]   I. McNab, Homopolar_Motor_and_Brush_Development_Studies, ADA436704.pdf (dtic.mil). Available from: https://apps.dtic.mil/sti/pdfs/ADA436704.pdf, accessed 29/12/24.

[13]   Microsoft PowerPoint – Navy_Motors-20190717 (cern.ch)]. Available from: https://indico.cern.ch/event/760666/contributions/3390601/attachments/1880202/3099643/Navy_Motors-20190715.pdf, accessed 29/12/24.

[14]   Superconductor Cables for Pulsed Dipole Magnets G.E. Gallagher-Daggitt, RHEL/M/A25, 1973.

[15]   PowerPoint Presentation (nist.gov). Available from: https://www.nist.gov/system/files/documents/pml/high_megawatt/Tomsic-DOE-NIST-Motor-Workshop-Sept-8-Mike-Tomsic-Hyper-Tech.pdf, accessed 29/12/24.

[16]   ABB motor sets world record in energy efficiency – saves half a million dollars. Available from: https://new.abb.com/news/detail/1789/ABB-motor-sets-world-record-in-energy-efficiency-saves-half-a-million-dollars, accessed 29/12/24.

[17]   F. Berg, J. Palmer, P. Miller and G. Dodds, *IEEE Trans. Appl. Supercond.* 27 (4), 1–7, Art no. 3600307 (2017), doi:10.1109/TASC.2017.2652319, accessed 29/12/24.

[18]   List of Large Brushless Motor Manufacturers + Summary Table – Tyto Robotics. Available from: https://www.tytorobotics.com/blogs/articles/brushless-motor-manufacturers-for-evtol-and-aviation, accessed 29/12/24.

[19]   Electric Propulsion System for Ship: Future in the Shipping? (marineinsight.com). Available from: https://www.marineinsight.com/marine-electrical/electric-propulsion-system-for-ship-does-it-have-a-future-in-the-shipping/, accessed 29/12/24.

[20]   Propulsion Systems (ricepropulsion.com). Available from: http://www.rice-propulsion.com/cartas/TNL49/Tnl49.htm, accessed 29/12/24.

[21]   High-Capacity and Efficiency Stirling Cycle Cryocooler. Available from: https://cryocooler.org/resources/Documents/C18/021.pdf, accessed 29/12/24.

[22]   R. Fuger, M. Arkadiy, K. John, D. Sercombe, and G. Ante. *Supercond. Sci. Technol.* 29 034001 (2016), doi:10.1088/0953-2048/29/3/034001.

[23]   Z. Huang, M. Zhang, W. Wang, and T. A. Coombs, *IEEE Trans. Appl. Supercond.* 24(3), Art. no. 4602605 (2014).

[24]   M. Baghdadi, H. S. Ruiz, and T. A. Coombs, *Sci. Rep.* 8(11), Art. No. 1342 (2018).

[25]   T Nakamura, K. Matsumura, T. Nishimura, *et al. Supercond. Sci. Technol.* 24, 015014 (2011).

[26]   T. Nakamura, K. Ikeda, G. Karashima, I. Yoshikawa, T. Itoh, and O. Terazawa, 2018 *International Symposium on Power Electronics, Electrical Drives, Automation and Motion*, 20–22 June 2018, Amalfi, Italy, pp. 22–1183, doi:10.1109/SPEEDAM.2018.8445204, accessed 29/12/24.

[27] W. J. Carr and C. E. Oberly, Filamentary YBCO conductors for AC applications, in *IEEE Transactions on Applied Superconductivity*, vol. 9, no. 2, 1475–1478.

[28] W. J. Carr Jr., *J. Appl. Phys.* vol. 45, no. 2, pp. 929–934, (1974) https://doi.org/10.1063/1.1663341, accessed 29/12/24.

[29] A. M. Campbell *Cryogenics* (1982).

[30] L. Hao, J. Hu, H. Wei, *et al.*, Transport AC losses in multiple-layer roebel tapes, in *IEEE Transactions on Applied Superconductivity*, vol. 33, no. 5, pp. 1–5, 2023, Art no. 4700805. doi: 10.1109/TASC.2023.3243567, accessed 29/12/24.

[31] Hongsoo Ha, Gwantae Kim, Hyunwoo Noh, Jaehun Lee, Seunghyun Moon and Sang-Soo Oh, *Supercond. Sci. Technol.* **33** 044007 (2020).

[32] A. Shah, Y. Öztürk, and H. Huang, *et al.*, A novel switch design for compact HTS flux pump, in *IEEE Transactions on Applied Superconductivity*, vol. 32, no. 6, pp. 1–5, Sept. 2022.

[33] A. M. Southerland and E. S. Brandler, *Neurology.* 88(14):1300–1301 (2017).

[34] C. Kilburg, J. Mcnally, A. de Havenon, P. Taussky, M. Y. Kalani, and M. Park, *Neurosurg Focus.* 42(4), E10 (2017), FOCUS16503. Review.

[35] M. Buller and J. P. Karis *Am. J. Neuroradiol.* 38(8), 1480–1485 (2017).

[36] Portable MRIs almost as effective as standard MRIs in detecting strokes – *ScienceDaily*. Available from: https://www.sciencedaily.com/releases/2022/04/220420151359.htm, accessed 29/12/24.

[37] L. J. M. van de Klundert and H. H. J. Ten Kate, *Cryogenics* 21(4), 195–206 (1981).

[38] D. L. Atherton, *Cryogenics* 7, 51 (1967).

[39] K. A. Hamilton, R. Badcock and D. Carnegie, *AIAA Propulsion and Energy*, Forum 10.2514/6.2020-3552, accessed 29/12/24.

[40] M. P. Risse, M. G. Aikele, S. G. Doettinger, R. P. Huebener, C. C. Tsuei, and M. Naito, *Phys. Rev. B* 55(15), 191 (1997).

[41] G. P. Mikitik and E. H. Brandt, *Phys. Rev. B* 64, Art. no. 092502 (2001).

[42] C. P. Bean, *Phys. Rev. Lett.* **8** (6), 250–253 (1962).

[43] Technavio, "High-temperature superconducting wires market trends [2023 report]." Available from: https://www.technavio.com/report/high-temperature-superconducting-wires-market-analysis, accessed 29/12/24.

[44] I. Patel, A. Shah, B. Shen, *et al.*, Stochastic optimisation and economic analysis of combined high temperature superconducting magnet and hydrogen energy storage system for smart grid applications, *Applied Energy*, vol. 341, 2023, 121070. doi:10.1016/j.apenergy.2023.121070, accessed 29/12/24.

[45] A. Molodyk, S. Samoilenkov, A. Markelov, *et al.*, *Sci. Rep.*, 11(1), 1–11 (2021), doi:10.1038/s41598-021-81559-z, accessed 29/12/24. 22, 25, 30, 31, 34, 44, 45.

[46] P. Fazilleau, X. Chaud, F. Debray, T. Lécrevisse, and J. Bin Song, *Cryogenics (Guildf)*, 106, 103053 (2020), doi:10.1016/j.cryogenics.2020.103053, accessed 29/12/24.

[47] European Magnet Field Laboratory (EMFL), "SuperEMFL – EMFL."

[48] J. H. Durrell, M. D. Ainslie, D. Zhou, *et al.*, *Supercond. Sci. Technol.*, vol. 31, 103501 (2018).

Chapter 5

Magnetic levitation technologies

Peter Berkelman[1]

5.1 Introduction

Magnetic levitation is a particularly compelling application of electromagnetism. Objects can be made to float, move and rotate under full control in any direction without any means of support, just like the anti-gravity technologies dreamed of in science fiction. But aside from the spectacle of stable controlled motion without any visible contact, magnetic levitation provides extraordinary distinctive advantages to application areas including transportation, parts handling, human–machine interaction and medicine due to the practical considerations of eliminating the friction, vibration and wear caused by contacts between moving objects.

The two principal requirements and challenges of magnetic levitation are to support the floating object and to stabilise its motion. All the forces and torques necessary to fulfill these two requirements are generated by electromagnetic interactions between various combinations of magnets and current-bearing electrical coils. The technical difficulties of magnetic levitation are that magnetic fields and the forces generated from them tend to decrease very rapidly with increasing distances relative to the size of the magnets or coils, and modelling of the forces generated is complex as these change with coil currents, time and the position and orientation of the levitated body.

The electromagnetic forces needed to support the mass of the levitated body vertically against the force of gravity depend on the dimensions and magnetisation of any permanent magnets used, and the size, shape and current of any electrified coils. Coil overheating is the major limitation on the levitation forces that can be generated with a given coil and magnet configuration, as the heat generated by a coil is proportional to the square of its electrical current.

Maintaining stability is critical in magnetic levitation systems. Due to the absence of frictional forces acting upon the motion of the levitated body, any net electromagnetic force that does not exactly compensate for the gravitational forces on the levitated body will instead produce motion away from its desired position and orientation. If these unwanted forces and motions are not corrected, they will

[1]University of Hawaii, Department of Mechanical Engineering, USA

increase over time without limit, resulting in an unstable system. Therefore, some means are needed to actively and continuously correct positioning errors in the magnetic levitation system. Typically, this stabilisation is achieved through the use of sensors and an active feedback control system, although somewhat more exotic methods are also possible such as gyroscopic stabilisation from a spinning body, using diamagnetic materials, or superconductivity.

It was in fact proven in 1842 by *Earnshaw's theorem* [1] that stable levitation cannot be maintained by any static arrangement of magnetic fields that can be generated by fixed magnets and coils with constant currents. In other words, no stable equilibrium point can exist with any static configuration of magnetic fields, and there will always exist some direction in either translation or rotation in which motion errors increase without limit. With only static fields, there is no point which is a local energy minimum, to which the levitated body will return if any small error or perturbation is present. Undesired motions of the levitated body due to unmodelled disturbances therefore need to be suppressed through feedback control.

With feedback control, sensors monitor the state of the system, and actuation forces are continuously modified according to any deviation from the desired state. When the feedback controller is properly designed, the total system errors can be minimised, and a naturally unstable system may be stabilised. To successfully achieve magnetic levitation, the motion of the levitated body must be stable in all directions in both translation and rotation. Under feedback control, the system can maintain a constant levitated position and orientation despite external perturbation disturbances, or follow a trajectory in which the desired state changes as a function of time. The trajectory can be defined as a changing position, orientation, or both in three dimensions.

The stable range of motion of a levitation system is a critically important parameter, depending on its application. This motion range may be different for each direction of translational and rotational motion, covering all six degrees of freedom of rigid-body motion. A suspension or vibration isolation system may only require stable levitation at a single fixed location and orientation, whereas a levitated rotational bearing must have unlimited rotation about a single axis, a planar motor needs a large range of translation in two directions, and a levitated train only needs translational motion in the direction of travel.

In magnetic levitation systems, optical sensors are typically used to measure the position of the levitated body for feedback control. Optical position sensor systems detect reflected or emitted light from markers on the levitated body and use the direction, time of flight or phase of the detected light to find the three-dimensional (3D) position and orientation of the levitated body by using combined data from multiple sensors, often by triangulation or other geometric techniques. The light may be generated from LED or laser sources and may be infrared or visible wavelengths, provided that other sources of light do not interfere with the fidelity of the sensors. Inertial or electromagnetic sensors may also be used. The controlled actuation signals then determine the electrical coil currents to generate forces and torques on the levitated body, according to a model of the interactions between the coils and magnets.

Using various combinations of coils and permanent magnets, electromagnetic forces and torques can be generated to lift and stabilise a rigid body. The capabilities and performance of magnetic levitation systems are continually increasing due to advances in sensors, magnetic materials, and computational capabilities.

The strongest permanent magnets presently available are made from alloys of rare earth elements rather than iron or ferrite. The strength of permanent magnet materials can be given by their *maximum energy product* $(BH)_{max}$, the maximum product of the B and H fields in the demagnetising curve of the material, measured in units of mega-Gauss-Oersteds or MGOe. Samarium-cobalt magnet alloys were primarily developed in the 1960s to 1970s, reaching a $(BH)_{max}$ of 33 MGOe. Neodymium magnets are made from an alloy of neodymium, iron and boron and have reached $(BH)_{max}$ levels of over 50 MGOe. The high strength of these magnets has led to their present adoption in electromagnetic devices and applications such as motors for electric vehicles and drones, headphones and speakers, disk drives and voice coils.

Comprehensive surveys of magnetic levitation systems and applications are given in *Rising Force: The Magic of Magnetic Levitation* [2] and *Magnetic Levitation: Maglev Technology and Applications* [3]. Unusual levitation systems which avoid using feedback control through instead adopting diamagnetic materials and spin stabilisation are described in [4].

The following section describes fundamental physical laws and models used in magnetic levitation. Various different methods and configurations that have been developed for magnetic levitation are surveyed in Section 5.3. Specific magnetic levitation systems developed in different application areas are described in Section 5.4. Section 5.5 concludes the chapter and describes present trends in magnetic levitation technology development.

5.2 Concepts and models

Electromagnetic forces are produced by interactions between permanent magnets, electric currents in wires, and any ferromagnetic materials that may be temporarily magnetised by external magnetic fields. If the spatial variations in magnetic fields are not excessively complex, the forces and torques produced in a given situation can be calculated directly from physical principles. With more complicated geometries, electromagnetic analysis software can be used. Calculation of magnetic fields and generated forces is particularly difficult when ferromagnetic materials such as iron are included, as the induced magnetisation of these materials may require iterative solution methods to calculate magnetic fields, forces and torques.

The magnetic field flux density at a point can be measured either in a single direction, as a total magnitude, or as a 3D vector by a magnetometer, otherwise referred to as a Gaussmeter. These instruments typically use the Hall effect or inductive sensing.

Electromagnetic forces are fundamentally produced from interactions between electric charges and magnetic fields according to Maxwell's equations. Direct application of Maxwell's equations is cumbersome for magnetic levitation modelling however, and simpler equations derived from Maxwell's are typically used.

If the magnetic fields in the system are constant in time or change slowly with respect to the rigid-body dynamics of the levitated object, then effects due to changing magnetic fields may be neglected and the system model is considered to be quasi-static. If magnetic fields are nearly constant with respect to position throughout the motion range of the levitated object, then the actuation model is also simplified considerably because the electromagnetic forces will not change when the object moves. Another simplification of the magnetic field is the magnetic dipole model, in which the shape and dimensions of a magnet or a cylindrical coil are neglected, and its effects are represented by a magnetic moment at a single point.

A magnetised body surrounded by a static magnetic field generates force **F** dependent on the gradient of the magnetic field and torque τ to align its magnetisation axis with that of the field as

$$\mathbf{F} = \int (\mathbf{M} \cdot \nabla)\mathbf{B}d\mathbf{V}, \tag{5.1}$$

$$\tau = \int (\mathbf{M} \times \mathbf{B})d\mathbf{V}, \tag{5.2}$$

for magnetisation **M** and external flux density **B**, integrated over the magnet volume **V**. If the coil dimensions are much larger than the magnet size, then **B** and its gradient are nearly uniform over the volume of the magnet and it is fairly straightforward to calculate **B** surrounding a cylindrical coil with constant current density according to the Biot–Savart law:

$$\mathbf{B} = \int \frac{\mu_0}{4\pi} \frac{I d\mathbf{l} \times \mathbf{r}}{r^3} \tag{5.3}$$

integrated over wire length **l**, vector **r** from the wire element $d\mathbf{l}$ to the point of evaluation of **B**, with current I and magnetic permeability of free space μ_0.

5.2.1   Lorentz forces

A current-bearing wire passing through a magnetic field produces a force **f** proportional to the current and field according to

$$\mathbf{f} = I \int d\mathbf{l} \times \mathbf{B}, \tag{5.4}$$

for a length of wire **l** carrying electrical current I in magnetic field **B**. As a result of the vector cross product in the equation, the force direction is perpendicular to both the current direction and the magnetic field direction, as shown in Figure 5.1.

If the magnetic field is constant along the length of the wire and throughout the range of motion of the magnet and wire, then the integral in (5.4) is a product and the calculation of the actuation force is straightforward and does not depend on position. Multiple lengths of wire are wound tightly together to form coils, so that

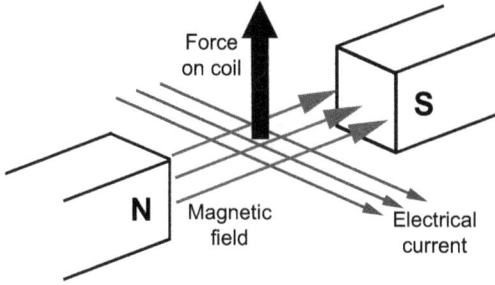

Figure 5.1 Lorentz force generated on coil wires in the magnetic field

the actuation force obtained from a given magnetic field and current is multiplied by the number of windings, and large forces may be obtained from relatively manageable levels of field strengths and current magnitudes. Besides avoiding dangerously high currents and voltages, the multiple windings in coils also reduce the power consumption and resistive heating in a magnetic levitation system, because the wire resistance and heat-dissipated power in Watts increase linearly with the number of coil windings, but increase as the square of the electrical current. Coil shapes are commonly rectangular, cylindrical or some combination of these such as 'racetrack' shaped coils.

5.2.2 Magnetic dipole model

The magnetic field lines in a plane containing the axis of a single cylindrical magnet with constant magnetisation throughout are shown in Figure 5.2.

To calculate the magnetic field around a single permanent magnet or cylindrical coil, it is common to represent its magnetic moment as emanating from a single point. In this way, all the dimensions and the magnetisation strength of a permanent magnet or electromagnet can be reduced to a single parameter, which determines the magnetic field model.

The magnetic field surrounding a magnetic dipole can be given as follows:

$$B_l = B_{lx}\widehat{i} + B_{ly}\widehat{j} + B_{lz}\widehat{k} = \frac{\mu_r \mu_0 M_T}{4\pi}\left(\frac{3(H_0 \cdot P)P}{R^5} - \frac{H_0}{R^3}\right), \tag{5.5}$$

where B_{lx}, B_{ly} and B_{lz} are the three components of the vector field B_l, with μ_r the relative permeability of the medium, μ_0 the magnetic permeability of air and M_T as a constant depending on the overall magnet strength. H_0 is the unit vector describing the orientation of the axis of magnetisation, P is the vector from the magnet center to the sensor and R is the distance to the magnet center.

This magnetic point dipole model does not account for the size and shape of the magnet and is most accurate at distances that are significantly larger than the dimensions of the magnet. If the magnet dimensions are comparable to the actuation distance, however, more complex modelling that accounts for the shape of the magnet is necessary to obtain accurate results.

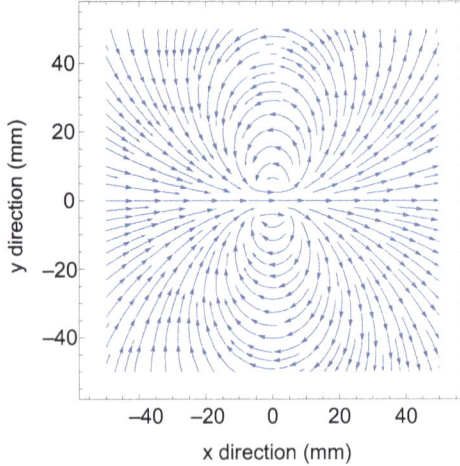

Figure 5.2   Magnetic field lines from the dipole magnet model

When a cylindrical coil is used to generate a magnetic field for attractive or repulsive actuation forces, an iron core in the center of the coil can greatly increase the generated forces due to the induced magnetisation of the iron. Modelling of the induced magnetisation of iron cores is considerably more difficult however, as the magnetisation depends on other surrounding magnetic fields as well, and the response time of the actuator may be reduced due to dynamic electromagnetic effects.

5.2.3   Induced currents

Whereas an electric current through a perpendicular magnetic field produces a Lorentz force perpendicular to both of these, in a symmetrical manner the relative motion between a conductor and a magnetic field produces an electromotive force, causing a current to flow in the conductor. Both of these properties contribute to the operation of electric motors and generators. The electromotive force, equivalent to electric voltage, is given by the equation

$$v_b = BLv, \tag{5.6}$$

where v_b is the induced voltage, B is the magnetic field, L is the length of the conductor and v is the perpendicular velocity component of the conductor moving through the magnetic field. If the conductor is an extended metal plate, then the electromotive force produces circular currents around the region where the magnetic field penetrates the conductor, referred to as eddy currents. The Lorentz force from the eddy currents acts against the motion which produces them.

The induced currents from the motion between conductors and magnetic fields may be significant in the operation of magnetic levitation systems in various ways, depending on the design and operation of the system. The forces from induced currents may be beneficial, as in the cases where eddy currents produce damping forces that oppose undesired motions, dissipating kinetic energy, vibration, and stabilising the motion of the levitated body. Induced currents can also produce vertical levitation forces from forward motion in electrodynamic suspension systems for maglev trains, as described in Section 5.4.1. Induced currents may also be detrimental to the performance of trajectory-following magnetic levitation systems because of resistive heating of the conductor and the generated forces adding drag to desired motion, wasting energy and producing viscous friction.

5.2.4 Rigid-body motion

The motion of a rigid body in three dimensions is always related to the forces and torques acting upon it by the Newton and Euler equations

$$\sum F = m\ddot{x}, \tag{5.7}$$

$$\sum \tau = I_c\dot{\omega} + \omega \times I_c\omega, \tag{5.8}$$

where $\sum F$ is the total of the forces acting on the center of mass of an object of mass m at position x, $\sum \tau$ is the total of the torques acting on the object, I_c is the 3-by-3 moment of inertia tensor of the object and ω is its angular velocity. When the velocities $\dot{x}$ and ω of a levitated object are low, the only significant forces acting on the object are gravity and the electromagnetic forces from the levitation system, and the only significant torques are electromagnetic torques. If velocities are sufficiently large, as in the case of a levitated vehicle, then aerodynamic forces and torques may also be significant. In either case, accurate modelling of all the electromagnetic and non-electromagnetic forces and torques allows the motion of the levitated body to be controlled. Furthermore, if the products of inertia, or the off-diagonal elements of the inertia tensor, are zero, or if the spatial angular velocity vector ω is negligible, then the $\omega \times I_c\omega$ term from equation can be neglected and the dynamics of each degree of freedom in translation and rotation about the center of mass are decoupled, and each degree freedom of translational and rotational motion may be controlled independently.

5.3 Methods

The operational parameters and capabilities of magnetic levitation systems depend on their actuation and sensing components, how they are arranged together in a particular configuration, and the control system used. Limitations on the capability of a magnetic levitation system may be due to any of these subsystems, and for optimal system design, it is necessary to determine which factors are primarily responsible for the limitations on the performance of a given system.

5.3.1 Feedback control

A magnetic levitation control system must always fulfill the requirements of lifting and stabilising the levitated object. If the object must follow a desired motion trajectory, then this function is also handled by the control system. Forces necessary for gravity compensation and producing the accelerations corresponding to desired motions can be calculated beforehand from the mass and trajectory, and similarly, the torques needed for a rotational trajectory can also be calculated beforehand to be generated by electromagnetic actuation. These forces and torques may be referred to as *feedforward* control terms, as they do not depend on the real-time sensed output of the system.

Feedback control terms stabilise the motion of the system and minimise its error. Without some form of feedback control, the instability of magnetic levitation systems would cause errors to increase over time without limit. A very common formulation of feedback control of a single output variable for a dynamic system is referred to as proportional-integral-derivative (PID) control. In PID control, actuation forces are calculated in real time from the sensed error of the system output as follows:

$$ f = K_P(x_d - x) + K_I \int (x_d - x)dt + K_D \frac{d}{dt}(x_d - x), \qquad (5.9) $$

where f is the actuation force, $x_d - x$ is the error signal which is the difference between the desired and actual output, and K_P, K_I and K_D are the three control gains which determine the action of the controller.

The proportional gain K_P produces an actuation force that is directly proportional to the error, and it is principally this action that drives the system output to its desired value, with a linear force that increases with distance in the same manner as a spring force. The derivative gain K_D produces an actuation force proportional to the derivative of the error signal. This control action is essentially a braking or damping force that acts to limit the speed or rate of change of the system output, dissipating energy and stabilising its motion. The integral gain K_I produces a control action proportional to the time integral of the error signal, so that a constant error produces a corrective action which increases over time. This integral gain is most effective when there are constant disturbances or inaccuracies in the system, which would otherwise produce constant errors. When the integral gain is properly tuned, the correcting force should increase to the point where it exactly cancels out the disturbance forces, thereby reducing steady-state errors to zero without otherwise destabilising the system.

The feedforward and feedback control terms can then be combined into a complete control system. If the dynamics of the levitation system are decoupled for each motion degree of freedom, then independent control can be implemented for each force and each torque acting on the system.

A considerable body of theory and methods have been developed to analyze the stability of feedback control of dynamic systems. Root locus analysis can be used for linear systems and Lyapunov theory can be applied to nonlinear systems. The non-linear variations in electromagnetic forces with position can be linearised in a small neighbourhood around a desired control setpoint to form a second-order dynamic model of each motion degree of freedom, which can be analysed and optimised in terms of performance measures such as stability, accuracy and response time.

Feedback control systems can be implemented as analog electronic systems, which act continuously on electrical systems, or as digital, discrete-time sampled-data systems. Analog control systems have the advantage of continuous action over time, but these may be susceptible to electrical noise in the signals. Digital control systems are not subject to signal noise issues, but they require analog-to-digital conversion of sensor signals to sampled data and digital-to-analog conversion of calculated forces to actuation signals. Sampling time and signal discretisation issues may also be significant in digital controllers if the sample time is not at least an order of magnitude faster than the required response time of the system or if the resolution of the analog-to-digital signal conversion is not sufficiently fine to accurately represent the size of the errors to be controlled.

5.3.2 *Sensing methods*

Depending on the size and mass of the levitation system, sensor resolutions of 100 μm or less and update rates of 1000 Hz or more may also be necessary to match the requirements of the feedback control system. Different types of sensors that can be used for contactless measurement of rigid-body position and orientation include optical, electromagnetic and inertial types. Ultrasonic sensing is another potential means of contactless position sensing; however, its application to magnetic levitation may be problematic due to limitations on its update rate and accuracy.

5.3.2.1 Optical sensing

Optical motion tracking systems use infrared emitters or reflectors fixed to the levitated object. Optical sensing is the most common means of measuring the position and orientation of the levitated object during operation for feedback control. It is free from interference due to electromagnetic fields, but it requires an unobstructed line of sight to the moving object.

Multiple sensors detect the directions to each emitter or reflector, and their 3D locations are determined by triangulation. Tracking of three or more locations is sufficient to determine the spatial orientation of the levitated rigid body.

To levitate objects at a tabletop scale, the position sensing update rate typically needs to be greater than 500 Hz and the tracking resolution less than 0.1 mm. Standard video cameras cannot deliver a sufficient frame rate, and the standard image resolution would require extreme magnification due to limited pixel size; however, specialised high-speed, high-resolution video cameras may have sufficient performance.

Specialised optical motion capture systems are available, which do have the necessary specifications. For example, Vantage V5 motion capture systems from Vicon can be configured for up to a maximum 2000 Hz frame rate with a precision higher than 0.1 mm. The OptoTrak Certus system from Northern Digital Inc. was also capable of a 1000 Hz motion tracking frame rate for a single object with three infrared markers, with precision higher than 0.1 mm. Combinations of position-sensing photodiodes and optical assemblies can be designed to deliver similar performance over small motion ranges without difficulty.

Optical position sensing is also possible based on the elapsed time or phase of a reflected beam, producing a measurement of distance rather than angle. LIDAR and time-of-flight sensors operate on these principles. Laser interferometry, in which the phases of light beams travelling different paths are compared, produces submicron precision measurements for use in fine motion magnetic levitation planar motor systems. Precisely smooth reflective surfaces are needed with a very small allowable range of rotation, typically a fraction of a degree.

5.3.2.2   Electromagnetic sensing

Capacitive sensing is possible over short distances, as the capacitance between two planar conductors in close proximity depends on their separation. It is necessary to ensure that induced currents from motion of conductors in magnetic fields do not interfere with the measurement of capacitance. Hall effect sensors detect fields by the deflection of charged particles in a conductor as they traverse a magnetic field, producing a measurable voltage signal. These sensors may measure field components in a single direction or on multiple axes and are commonly used in servomotors and other electrical machines. In magnetic levitation, the positions and orientations of moving magnets are calculated from their detected fields measured by multiple sensors.

Magneto-inductive sensors detect induced current in coils and can deliver much higher precision measurement of magnetic fields than Hall effect sensors [5]. A three-axis precision sensor system is available on a compact circuit board as the RM3100 from PNI Corp.

The principal advantage of electromagnetic position sensing compared to optical methods is that an unobstructed line of sight is not necessary between the sensor and emitter. The disadvantage is that accurate modelling and compensation of the fields produced by coil currents is needed to detect fields due to moving magnets and to determine their positions.

5.3.2.3   Inertial sensing

An inertial measurement unit (IMU) senses the six degrees of freedom of rigid body motion using a three-axis accelerometer and gyroscope. Acceleration signals must be numerically integrated twice to find velocity and position, and the angular velocity measurement obtained from the gyroscope must be integrated once to obtain a measurement of rigid-body position and orientation. Because these measurements are not given with respect to a fixed reference frame, small errors in the derivative signals will accumulate over time and produce drift in the position and orientation measurements. For a limited amount of time, changes in position and orientation relative to a known starting point may be calculated with sufficient accuracy, so that the measurements of linear and angular velocity are highly useful for stabilisation control. If the actual acceleration of the rigid body is low, then the acceleration vector given by the accelerometer actually measures the direction of gravity, so that an IMU may also function as an inclinometer under some conditions.

IMUs are typically used in drones because they can be produced at low weight and low cost, and they do not have any limitation on their range of motion. Small IMU sensors are available on circuit boards and easily integrated with control processors. Large, costly and very precise IMU systems are used for military applications.

5.3.3 Modelling

When the design and operating parameters of a magnetic levitation system are such that changes in magnetic fields are very small over the range of motion, or if distances between magnets and coils are sufficiently large to obtain accurate results using magnet dipole models as described in Section 5.2, then the calculation of forces and torques from coil currents for use in feedforward control and stability analysis is straightforward, using approximate equations directly in the calculations. Under conditions where dipole or constant field approximations do not provide accurate results, more precise models must be used.

For more accurate modelling of electromagnetic fields, forces and torques, electromagnetic modelling software can be used. These software packages are based on finite element and boundary analysis, in which objects and volumes are subdivided into surface or volumetric meshes, and extensive calculations are carried out on the elements and nodes of the meshes. These calculations are computationally intensive and not suitable for real-time operation; however, precalculated lookup tables of force and torque generation dependent on position and orientation may be stored for use during operation.

COMSOL multiphysics finite element analysis software includes electromagnetic modelling capabilities. *Ansys Maxwell* (formerly Ansoft) is specifically developed for 3D low-frequency EM field simulation and is used in many electromagnetic analysis applications. *Radia* [6] is a free software package, which was developed by the European Synchrotron Radiation Facility for analysis of magnet systems. Programming interfaces for *Radia* are available for Python or Mathematica.

5.3.4 Coil and magnet configurations

Several common configurations for the coils and magnets in magnetic levitation systems are listed below, in order from simple to more complex. Schematic representations of these sample configurations are shown in Figures 5.3–5.5. Configurations for maglev trains are given in Section 5.4.1, as the lift, guidance and propulsion functions may be considered separately in these systems.

5.3.4.1 Single coil suspension

It is possible to levitate a single small magnet with a single coil and sensor, provided that the magnet is suspended underneath the coil by an attractive force. A steel ball may also be suspended, as magnetisation will be induced by the coil, producing the attractive force. In this simple configuration the dynamics of the suspended rigid body are similar to that of a pendulum, where sideways swinging motions may be present, but these degrees of freedom are stable as the force of gravity always draws the magnet to a centered position underneath the coil. Position feedback can be provided by any sensor that produces a signal dependent on the magnet position, such as a light beam and detector, a capacitor or a Hall effect sensor.

The steady-state height of the suspended ball or magnet may be changed by adjusting the position setpoint of the controller, but the horizontal position and vertical orientation of the magnet are fixed. A variation of this configuration has

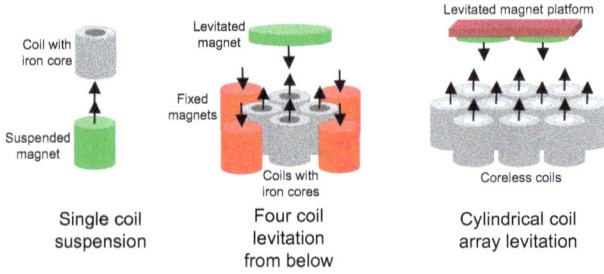

Figure 5.3   Coil and magnet configurations for magnetic levitation systems

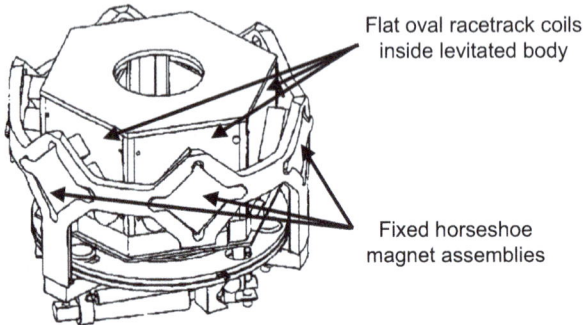

Figure 5.4   Lorentz levitation robot wrist

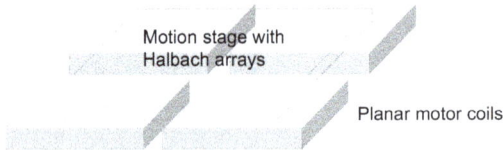

Figure 5.5   Planar motor levitation system

been developed by Khamesee *et al.* using four coils connected by a cross-shaped pole piece, so that the differences in current between the opposite coils in each pair change the position of the suspended magnet equilibrium point in the two horizontal directions [7].

5.3.4.2   Four coil levitation from below

A four-coil configuration exists, which is commonly available, low cost, simple and results in levitation from below, with the magnet above the coils. The notable features of this configuration are that only four coils and two Hall effect sensors are needed to levitate a single magnet from underneath. Simple proportional-derivative (PD) controllers are sufficient to stabilise both translation and rotation in both planar directions in a small motion region. Translation in the vertical direction is stable and rotation about the vertical axis is neutrally stable and uncontrolled, allowing a disk magnet to rotate freely.

In this levitation configuration, an outer ring of permanent magnets provides passive lifting forces to support the levitated magnet in the center. The polarities of the fixed permanent magnet ring and the levitated magnet are in the same direction, and the induced magnetisation of the iron cores of the central coils in the opposite direction provides the lifting force. The four coils are connected together in two pairs, and the windings of the two coils in each pair have opposite polarities, so that a single current through a coil pair generates antagonistic forces to push and pull the levitated magnet in one planar direction, and rotate the magnet about one planar axis. Two Hall effect sensors at the center of the levitation system detect off-center components of the magnetic field from the levitated magnet in the two planar directions, so that the feedback signal from each Hall effect sensor determines the current through its corresponding pair of electromagnetic coils according to its digital PD controller.

This levitation configuration was patented in [8] in 1992 and the patent holder Levitation Arts Inc. has made it available for a wide variety of decorative displays which are commonly available. A variation on this configuration is produced by Crealev BV, which uses an upward facing optical sensor at the center of the device, which tracks a reflective region on the bottom surface of the levitated platform in the place of Hall effect sensors.

5.3.4.3 Flat coil planar motor levitation

Kim and Trumper developed a magnetic levitation motion stage based on planar motors [9], with a planar motion range of 50×50 mm, a vertical motion range of 400 µm and milliradian scale rotation in all directions. Four sets of flat stator windings are used to generate levitation and planar forces against magnets in the base of the levitated stage. The levitated magnets are arranged in a planar Halbach array [10], a repeating pattern of permanent magnets magnetised in different directions, used to concentrate magnetic fields on one side of the magnets, as in Figure 5.5. The device is capable of 5 nm precision motions in all directions. Capacitance probes are used for vertical position sensing, and laser interferometer systems are used for planar position sensing.

Planar Motor Inc. and Beckhoff Automation LLC have commercialised levitated planar motor systems with a novel design that allows multiple stators to be tiled together so that the planar motion range can be extended to any length [11]. Large-scale rotation about the vertical axis and control of multiple levitated movers on a single stator is possible under limited conditions, as described in the patent [12].

5.3.4.4 Lorentz levitation

Hollis and Salcudean invented a magnetic levitation method that uses Lorentz force actuation [13]. Each actuator consists of a pair of fixed horseshoe-shaped magnets located on either side of a thin flat racetrack coil so that magnetic flux in opposite directions is concentrated on areas on opposite sides of each coil, as in Figure 5.4. The two coil areas that pass through high magnetic fields produce Lorentz forces in the same direction, because both the direction of the field and the direction of the current are reversed between the two areas, so that the resulting force vectors from their cross products are in the same direction and add their magnitudes together directly.

To minimise the levitated mass, the magnet assemblies are fixed, and the coils are embedded in a thin shell on the levitated body. Six of these Lorentz force actuators, arranged in different locations and directions and acting in parallel, are sufficient to generate combinations of forces and torques in all directions so that all six degrees of freedom of the levitated rigid body motion can be stabilised.

Given the six coil locations and orientations and their current-to-force proportional actuation constants, it is possible to calculate a 6 × 6 element square transformation matrix, which maps the six coil currents to a six-element vector of force and torque components in all directions. The inverse of this matrix is then used by the control system to calculate the six coil currents needed to generate the feedback control forces at each sensor and control update.

Position and orientation sensing for feedback control is implemented using three infrared LED markers on the levitated shell, facing three planar position sensing photodiodes on the fixed base. An early application of Lorentz levitation was as a fine motion robot wrist referred to as the *Magic Wrist* [14], with 10 mm translation and 10° rotation ranges. A later development of Lorentz force levitation used a novel coil winding shape and two layers of coils on a hemispherical shell to realise 50 mm translation and 60° rotation ranges in all directions [15].

5.3.4.5   Cylindrical coil array

Platforms containing any number of magnets can be levitated using planar arrays of cylindrical coils packed closely together, as in Figure 5.3. Since each coil has a different position relative to each magnet, each one generates a different combination of forces and torques from a given current. With a sufficient number of coils, it is possible to generate a total force and torque vector in any direction as needed, if the correct combination of currents is supplied to the coils at each update of the feedback control system.

To calculate these coil currents at each control update, a precise actuation model is needed which gives the forces and torques generated between a single coil and a single magnet, according to their relative position and orientation and the current supplied to the coil. A numerical actuation model may be obtained either from measurements or from electromagnetic modelling software and stored in a large multidimensional lookup table indexed by position and orientation. In practice, a lookup table indexed to the closest mm in position and the closest 10° in orientation is sufficient for stable levitation. Using the lookup table model, a transformation matrix between the vector of all coil currents and the total force and torque vector generated on the levitated body can be calculated at each update by combining all the forces and torques between each magnet and each coil according to its current. The transformation matrix must be inverted to find the currents needed for control forces and torques. With redundant actuation due to more coil currents than the six-element force and torque vectors, the matrix pseudoinverse is used to provide the least-squares optimal set of currents with the lowest total power requirements.

Position and orientation feedback is given by an optical motion tracking system with infrared LED markers. The planar motion can be extended in either direction by adding coils to the array. A 2011 levitation system implementation using 16 coils to levitate a platform with 2 disk magnets demonstrated a motion range of $80 \times 80 \times 25$ mm [16]. A similar implementation demonstrated an unlimited rotation range in all directions with 27 coils and a levitated platform containing six disk magnets arranged on the faces of a cube [17]. An implementation using square coils demonstrated a 100×40 mm horizontal range [18].

5.3.5   *Exotic levitation methods*

Under certain special conditions, a feedback control system is not necessary for stable magnetic levitation due to specific physical phenomena such as diamagnetism, superconduction and gyroscopic stabilisation. Appropriately shaped magnets and conductors can also demonstrate stable levitation by electromagnetic induction using high-frequency alternating current.

The *Levitron* is a top that can levitate stably above a permanent magnet due to gyroscopic or spin stabilisation [19] and is the topic of several related patents. It is stable within a limited range of spin rates, so its levitation is for a limited time as the spin rate decreases unless the spin can be externally driven to maintain a constant angular velocity.

Diamagnetic materials such as water and other organic materials are repelled by magnetic fields and may therefore be levitated by constant fields with no feedback control stabilisation. The effect is typically weak and requires very high fields from an electromagnet for levitation of small objects such as a frog, as reported in [4]. Bismuth and graphite are more strongly diamagnetic, and pyrolitic graphite in particular can be levitated using neodymium permanent magnets.

The effects of diamagnetism are several orders of magnitude greater in superconducting materials, which can expel magnetic fields from their interior by the Meissner effect. Typical metals require temperatures near absolute zero to become superconductive, but compound materials have been found that show superconductivity above the temperature of liquid nitrogen at 77 K. The phenomenon of flux pinning in these superconducting compounds enhances stability and permits levitation in any direction.

5.4   Applications

The non-contact support and controlled stable motion made possible by magnetic levitation provide benefits in application areas including transportation, manufacturing, engineering testing and design, medicine and human–computer interaction. The advantages of eliminating friction, wear, impact, vibration and complex mechanisms have been applied at motion and mass scales ranging from 100 m, 100 kt objects moving at hundreds of km/hour in transportation, to mg objects moving nm distances for medicine and microfabrication.

5.4.1   Transportation

The most commonly known application area of magnetic levitation and its most active area of research and development is for transportation, otherwise known as maglev trains [20]. The advantages of maglev trains over conventional wheel-on-track trains due to their non-contact propulsion and guidance include a smooth ride and lack of wear, enabling operation at high speeds. Several systems have been developed but at present only a few systems have ever been deployed for public use, and in limited areas only. The obstacles to wider adoption of maglev trains may include high costs, incompatibility with existing infrastructure and the difficulty of switching between tracks.

For maglev trains, the dynamics and control of levitation, guidance and propulsion may be considered separately. Here, the levitation forces are vertical, guidance forces are perpendicular to the direction of travel and propulsion forces are in the direction of travel. The lifting and guidance forces act to keep vehicle cars lifted and centered within a small range of horizontal and vertical separation distance from the track. Propulsive forces are produced by travelling waves of alternating field directions between the vehicle and track electromagnets. This propulsion essentially operates as a linear motor, which may be considered as a conventional multiple pole electric motor in which the rotor and stator are 'unrolled' into a continuous linear path.

Whereas levitated strains eliminate the friction of wheels on tracks, vacuum tube train systems take the next step of eliminating all drag from air resistance by proposing to run trains inside tunnels from which all air has been evacuated. The Swissmetro project is in development in Switzerland and claims political support, but no definite plans exist for constructing a working system. A *hyperloop* is a similar concept which was promoted by Elon Musk in 2013. A test track was built by SpaceX in 2015 and the *Hyperloop One* company was established in 2014. Experimental uncrewed pods have been developed by student teams including Massachusetts Institute of Technology, Delft University and the Technical University of Munich, reaching a maximum speed of 463 km/h in 2019.

The two most common approaches to magnetic levitation train systems are generally referred to as electromagnetic suspension (EMS) and electrodynamic suspension (EDS). The principal differences between these two approaches are given below.

5.4.1.1   Electromagnetic suspension

In EMS systems, the bottom part of each train car wraps around each rail and extends beneath it, as shown in Figure 5.6(a). This structure enables lifting forces to be generated by electromagnetic attraction between the rails and the underside of each train car. The separation distance of levitation is typically approximately 10 mm. In some cases, sufficient guidance forces are produced by the self-centering action of the attractive lifting force. Otherwise, additional electromagnetic forces for guidance may be generated between the sides of the rail and the inside surface wrapped around the rail, also shown in Figure 5.6(a). Propulsive wave forces

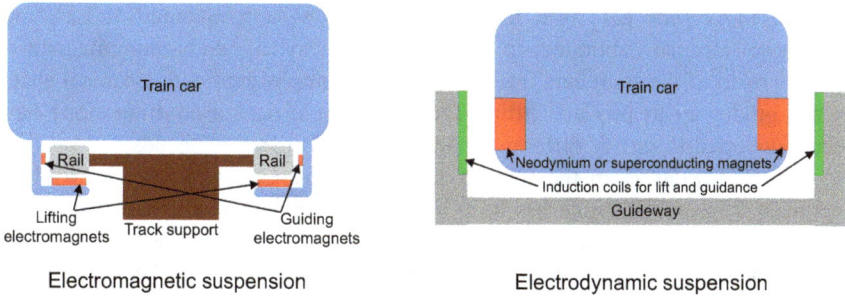

Electromagnetic suspension   Electrodynamic suspension

Figure 5.6   (a) Electromagnetic and (b) electrodynamic train suspension examples

operate independently of lifting and guiding forces, allowing the train to be levitated at very low speeds or standing still.

The first conceptions of EMS trains emerged in the early 1900s, including patents by Hermann Kemper in the 1930s. At present, the most successful EMS system has been the *Transrapid* monorail. A test track was initially built in Germany in the 1980s and a commercial public system was completed in 2002 as a high-speed connector between the city of Shanghai, China and Shanghai Pudong International Airport. This line covers a distance of 30.5 km in 7–8 minutes, reaching a maximum speed of 431 km/hour which is the world's fastest in commercial operation. Other EMS systems which have achieved commercial operation at low or medium speeds other than temporary exhibitions include the following:

Location	Length	Speed	Years of operation
Birmingham Airport, UK	600 m	Low speed	1984–95
Berlin, Germany	1.6 km	80 km/h	1989–91
Nagoya, Japan	8.9 km	100 km/h	2005–Present
Incheon Airport, Korea	6.1 km	80 km/h	2016–Present
Changsha, China	18.5 km	100 km/h	2016–Present
Beijing, China	10.2 km	100 km/h	2017–Present

5.4.1.2   Electrodynamic suspension

The defining feature of EDS systems is that the motion of the train induces currents on the track, which produce the needed lifting forces. The magnetic fields involved are much greater than those involved in EMS, levitation forces are repulsive rather than attractive, and the levitation distance can be on the order of 100 mm. Since forward motion is needed for levitation, the train lowers itself onto wheels as it slows to a stop and gradually lifts into the air as it accelerates. Due to the induced currents and the geometry of the coils and magnets on the track and train, it is possible to design an EDS system with passive stability with respect to the levitation height and guidance directions. The general arrangement of magnets and coils for EDS maglev train systems is shown in Figure 5.6(b).

The *Inductrack* [21] is a fully passive EDS system invented at Lawrence Livermore National Laboratory in 2005. Halbach arrays of neodymium magnets on the underside of the train cars interact with either unpowered coils or metal sheets on the guideway to produce lifting forces. Multiple design configurations have been developed to satisfy different combinations of operating speeds and loads.

The *SCMaglev* from the Central Japan Railway Company and the Railway Technical Research Institute has been in development since 1972 and is presently the most advanced EDS system in terms of technology and implementation. The train contains large superconducting electromagnets, and coils are embedded into the side walls of the guideway. A 7-km test track was constructed in 1977 and an 18-km track from 1990 to 1997. A maximum train speed of 603 km/h was reached on the most recent test track in 2015.

In 2011, the SCMaglev was selected to operate on the Chuo Shinkansen express line currently under construction between Tokyo and Nagoya and is planned to connect the two cities in 40 minutes with a 505-km/h operating speed over a 286-km route, mostly through tunnels. The line is planned to reach Nagoya in 2027 and extend to Osaka in 2037.

5.4.2   Precision material handling and motion control

The semiconductor manufacturing industry requires submicron precision motion of silicon wafers in clean room conditions for photolithography. Planar motor magnetic levitation systems [9] can meet these requirements when sufficiently precise position sensing is used, as the noncontact electromagnetic actuation does not produce any dust to contaminate the clean room conditions. In these systems, flat rectangular coils are used, the gap between coils and magnets is very small and significant motion ranges are only required in planar directions. Laser interferometry provides position sensing with the needed precision. Halbach arrays of magnets concentrate magnetic fields on the side of the moving plate where the fixed coils are located.

Magnetic levitation bearings [22] eliminate friction on a rotating shaft. This feature is particularly desirable for high-speed precision rotation systems such as flywheel energy storage.

5.4.3   Wind tunnel suspension

Magnetic levitation systems are useful for the support of aerodynamic models in wind tunnels because there need be no support present, which would otherwise interfere with airflow and accurate measurements. Large strong magnets and coils are needed because the gaps between the fixed coils and levitated magnet may be 25 cm or greater. A robust control system is needed to reject the force and torque disturbances from high-velocity airflows while maintaining the levitated body in a fixed position and orientation.

The other advantage of magnetic levitation support of models in wind tunnels is that aerodynamic forces and torques can be calculated very accurately, provided that an accurate actuation force model is known. According to free-body analysis, levitation forces, gravitational forces, and aerodynamic forces must sum to zero while the levitated body is motionless. Therefore, it is sufficient to know the weight of the model and the forces and torques generated by the controller to find the aerodynamic forces.

Wind tunnel magnetic suspension systems as described have been developed by NASA [23] and by the National Aerospace Laboratory of Japan [24].

5.4.4 Medical applications

Controlled motion of magnetic capsules inside the human body has a high potential benefit for medical care. Autonomous capsules moving noninvasively inside the body could perform functions including drug delivery, biopsy and diagnostic sensing, which would otherwise require invasive surgery at much higher cost, risk, trauma and recovery time to the patient.

In these applications, stable levitation is not needed, as the magnetic capsules remain in contact and are supported by bodily tissues or fluids, and therefore unstable dynamics do not need to be stabilised. The challenges of electromagnetic motion control for magnet capsules inside the body are from the large distances between coils and magnets, large required motion ranges, non-optical position sensing, and modelling of interaction with bodily tissues. Notable systems include the medical magnetic navigation systems developed by Stereotaxis Inc. [25] and the *Octomag* [26] system for controlled manipulation of magnetic microbot capsules inside the eye.

5.4.5 Haptic interaction

Haptic interaction refers to human interaction with physical, remote or virtual objects through the tactile, kinesthetic and proprioceptive aspects of the sense of touch. Magnetic levitation devices are particularly well suited as haptic interfaces for computer-mediated interaction with simulated or remote environments because the rigid-body fine motion dynamics of a grasped object can be reproduced in all six degrees of freedom in translation and rotation with controlled force and position precision and frequency response bandwidths similar to what is perceivable by the senses of the human hand. The high fidelity of haptic interaction with a magnetic levitation device is in fact due to the non-contact actuation and sensing, which eliminates friction. With a magnetic levitation haptic interface device, a levitated handle can realistically reproduce the motions and sensations of a handheld tool or utensil while it is grasped and manipulated by the hand of a user.

Lorentz force magnetic levitation devices were first invented and developed for haptic interaction by Hollis and Salcudean [27]. A desktop-sized magnetic levitation haptic interface device with a fingertip range of motion was developed and integrated with a graphics workstation to enable 3D graphic and haptic interaction with real-time dynamic simulated environments by Berkelman [28]. A similar device was commercialised by Hollis as the *Maglev 200* magnetic levitation haptic interface from Butterfly Haptics LLC.

The planar coil array magnetic levitation system from [16] was also developed as a haptic interface device. A novel advantage of the planar coil array is that a thin flat display can be placed between the coils and the levitated handle grasped by the user, so that the haptic interaction and the graphical display of the simulated environment are physically colocated [29].

5.5   Potential future directions

Future developments in magnetic levitation are difficult to predict accurately, but continuing research and development in areas including magnetic materials, sensing and control may lead to dramatic new applications in transportation, medicine and haptic interaction. When these remaining technical challenges are overcome, it is only necessary to adapt new technologies to the correct scales of distance and force.

5.5.1   Superconducting materials

All magnetic levitation systems are limited in terms of their force, levitation distance, motion speed and range by the magnitude of the magnetic fields which are used. In practice the magnetic fields are produced either by permanent magnets or by electric currents. The development of neodymium alloy (NdFeB) magnets in the 1980s produced the strongest permanent magnets available today, with magnetisation B_r up to 1.4 T and energy density BH_{max} up to 55 MGOe.

To generate more powerful magnetic fields, electrical currents through coils are needed. These currents and the generated fields are limited by power supply capabilities and resistive heating of the coil wiring. Therefore, superconducting materials for magnetic levitation coil windings have the potential to dramatically impact the capabilities of magnetic levitation systems. In fact, the *Chuo Shinkansen* maglev rail system presently under construction between Tokyo and Nagoya, Japan, uses superconducting electromagnets, which must be cooled to −269 °C using liquid helium.

If room temperature superconducting materials are developed, the impact on maglev transportation and other large-scale applications will be considerable. Room temperature superconducting magnets would eliminate the power requirements for electromagnetic actuation as well as cooling requirements, and the generated magnetic field would be limited only by the current limitation of the superconducting material. Yet although room temperature superconducting materials have been announced several times by different research groups, these results have not been validated at present.

5.5.2   Airless tunnels

Maglev train systems eliminate the rolling resistance of wheels on tracks, leaving air resistance on high speed motion to be overcome. If maglev transportation is run through airless, evacuated tunnels, then ground transportation will be able to operate at speeds and efficiencies that have never been achieved before. The concept is clear and there is no technological obstacle to realising maglev trains in airless tunnels, but economic factors still stand in the way, as underground tunneling and maintaining vacuum conditions for long-distance travel require significant infrastructure investment.

When transportation systems on the moon or other bodies with no atmosphere are considered, magnetically levitation becomes increasingly advantageous, as sealed tunnels are not needed to obtain the benefits of completely frictionless motion. Substantial investment for the construction of guideways and installation of magnets and coils would be necessary to obtain these possible benefits; however, so the possible application of magnetic levitation on the moon will be limited for the foreseeable until significant lunar infrastructure is already in place [30].

Space launch systems are a particularly attractive potential application area for maglev propulsion through airless tubes, as the acceleration and velocity needed to reach earth orbit could be produced at greatly reduced cost as compared to conventional rocket propulsion. The *StarTram* has been proposed as a maglev launch system, with its preliminary design and analysis described in [31,32]. The concept is to use superconducting maglev train technology to accelerate a launch vehicle through an evacuated tube, which is horizontal at ground level and slopes upwards so that the vehicle is released at an altitude of 4000 m or more. The vehicle would then ascend to orbit either with additional rocket thrust or simply by coasting through the thin atmosphere. The advantages of this launch system are that the vehicle does not have to carry the large amount of fuel needed for rockets, and much of the atmospheric drag during launch is avoided by release from the tube at high altitude. It is claimed that such a system would have a launch cost of $50 per kg of payload, as compared to $10,000 per kg using rockets.

5.5.3 Medical capsules

Accurate motion control and manipulation of miniature wireless capsules inside the human body would be a significant step forward, as many internal medical procedures would become completely noninvasive, including biopsy, targeted drug delivery and removing or destroying tumors. Although systems presently exist for generating electromagnetic forces on magnetic capsules in the body, these have severely limited positioning accuracy and responsiveness, and typically require the use of large external robotic arms to manipulate coils or permanent magnets. The technological challenges that remain to be overcome are in accurate and responsive sensing of the capsule position and orientation so that precise feedback control can be implemented using electromagnetic forces and torques. These sensing methods are a promising area for further development.

5.6 Discussion and conclusions

Although magnetic levitation systems have been developed and analysed for many decades leading to the present time, it is still an active area of technological development. The contributing technologies of high energy product magnetic materials and higher temperature superconducting materials for force and torque generation, improvements in sensor precision and response time for motion accuracy and speed, and high-speed computational capabilities for analysis, optimisation, and real-time control continue to have a significant impact on improving the performance and broadening the application of magnetic levitation systems.

Maglev trains are likely to become the most impactful application of magnetic levitation in the near future. After decades of short-lived, discontinued train systems, more newly built maglev train systems have begun commercial operation from 2016. The most ambitious commercial application in terms of cost, length, speed and technology is the *SCMaglev*, with a record maximum speed of 603 km/h and planned for use on the 286 km Chuo Shinkansen route under construction between Tokyo and Nagoya by 2027.

References

[1]   S. Earnshaw, "On the nature of the molecular forces which regulate the constitution of the luminiferous ether," *Transactions of the Cambridge Philosophical Society*, vol. 7, pp. 97–112, 1842.
[2]   J. D. Livingston, *Rising Force: The Magic of Magnetic Levitation*. Cambridge, MA: Harvard University Press, 2011.
[3]   H.-S. Han and D.-S. Kim, *Magnetic Levitation: Maglev Technology and Applications*. Berlin: Springer, 2016.
[4]   M. V. Berry and A. K. Geim, "Of flying frogs and Levitrons," *European Journal of Physics*, vol. 18, no. 4, p. 307, 1997.
[5]   S. Tumanski, "Induction coil sensors: A review," *Measurement Science and Technology*, vol. 18, no. 3, p. R31, 2007.
[6]   O. Chubar, P. Elleaume, and J. Chavanne, "A three-dimensional magnetostatics computer code for insertion devices," *Journal of Synchrotron Radiation*, vol. 5, pp. 481–484, 1998.
[7]   M. B. Khamesee, N. Kato, Y. Nomura, and T. Nakamura, "Design and control of a microrobotic system using magnetic levitation," *IEEE/ASME Transactions on Mechatronics*, vol. 7, no. 1, pp. 1–14, 2002.
[8]   L. Whitehead, "Levitation system with permanent magnets and coils," *US Patent* 5168183, 1992.
[9]   W.-J. Kim and D. Trumper, "High-precision magnetic levitation stage for photolithography," *Precision Engineering*, vol. 22, pp. 66–77, 1998.
[10] K. Halbach, "Design of permanent multipole magnets with oriented rare earth cobalt material," *Nuclear Instruments and Methods*, vol. 169, no. 1, pp. 1–10, 1980.
[11] X. Lu and I. ur-rab Usman, "6D direct-drive technology for planar motion stages," *CIRP Annals – Manufacturing Technology*, vol. 61, no. 1, pp. 359–362, 2012.
[12] X. Lu and I. ur-rab Usman, "Displacement devices and methods for fabrication, use and control of same," International Patent WO2 013 059 934A1, 2012.

[13] R. L. Hollis, "Six DOF magnetically levitated fine motion robot wrist with programmable compliance," U. S. Patent No. 4,874,998, October 1989.

[14] S.-R. Oh, R. L. Hollis, and S. E. Salcudean, "Precision assembly with a magnetically levitated wrist," in *IEEE International Conference on Robotics and Automation*, Atlanta, May 1993, pp. 127–134.

[15] P. Berkelman, "A novel coil configuration to extend the motion range of Lorentz force magnetic levitation devices for haptic interaction," in *IEEE/RSJ International Conference on Intelligent Robots and Systems*, San Diego, October 2007.

[16] P. Berkelman and M. Dzadovsky, "Magnetic levitation over large translation and rotation ranges in all directions," *IEEE/ASME Transactions on Mechatronics*, vol. 18, no. 1, pp. 44–52, 2013.

[17] M. Miyasaka and P. Berkelman, "Magnetic levitation with unlimited omnidirectional rotation range," *Mechatronics*, vol. 24, no. 3, pp. 252–264, 2014.

[18] X. Zhang, C. Trakarnchaiyo, H. Zhang, and M. B. Khamesee, "Magtable: A tabletop system for 6-DOF large range and completely contactless operation using magnetic levitation," *Mechatronics*, vol. 77, p. 102600, 2021.

[19] T. Jones, M. Washizu, and R. Gans, "Simple theory for the Levitron®," *Journal of Applied Physics*, vol. 82, no. 2, pp. 883–888, 1997.

[20] H.-W. Lee, K.-C. Kim, and J. Lee, "Review of maglev train technologies," *IEEE Transactions on Magnetics*, vol. 42, no. 7, pp. 1917–1925, 2006.

[21] R. Post and D. Ryutov, "The inductrack: A simpler approach to magnetic levitation," *IEEE Transactions on Applied Superconductivity*, vol. 10, no. 1, pp. 901–904, 2000.

[22] G. Schweitzer, H. Bleuler, and A. Traxler, *Active Magnetic Bearings – Basics, Properties, and Applications*. Zurich: Hochschulverlag AG, 1994.

[23] N. J. Groom and C. P. Britcher, "A description of a laboratory model magnetic suspension testfixture with large angular capability," in *IEEE Conference on Control Applications*, Dayton, September 1992, pp. 454–459.

[24] H. Sawada, S. Suda, and T. Kunimasu, "NAL 60cm magnetic suspension and balance system," in *Congress of International Council of the Aeronautical Sciences*, August 2004, pp. 2004-3.1.2.

[25] D. Meeker, E. Maslen, R. Ritter, and F. Creighton, "Optimal realization of arbitrary forces in a magnetic stereotaxis system," *IEEE Transactions on Magnetics*, vol. 32, no. 2, pp. 320–328, 1996.

[26] M. P. Kummer, J. J. Abbott, B. E. Kratochvil, R. Borer, A. Sengul, and B. J. Nelson, "OctoMag: An electromagnetic system for 5-DOF wireless micromanipulation," *IEEE Transactions on Robotics*, vol. 26, no. 6, pp. 1006–1017, 2010.

[27] R. L. Hollis and S. E. Salcudean, "Lorentz levitation technology: A new approach to fine motion robotics, teleoperation, haptic interfaces, and vibration isolation," in *Proceedings of the 6th International Symposium on Robotics Research*, Hidden Valley, PA, October 1993.

[28] P. J. Berkelman and R. L. Hollis, "Lorentz magnetic levitation for haptic interaction: Device design, function, and integration with simulated

environments," *International Journal of Robotics Research*, vol. 9, no. 7, pp. 644–667, 2000.

[29]   P. Berkelman, M. Miyasaka, and J. Anderson, "Co-located 3D graphic and haptic display using electromagnetic levitation," in *IEEE Haptics Symposium*, Vancouver, March 2012, pp. 77–81.

[30]   U. Apel, "Comparison of alternative concepts for lunar surface transportation," *Acta Astronautica*, vol. 17, no. 4, pp. 445–456, 1988.

[31]   J. R. Powell, G. Maise, J. Paniagua, and J. D. Rather, "Startram: A new approach for low-cost earth-to-orbit transport," in *2001 IEEE Aerospace Conference Proceedings*, vol. 5, 2001, pp. 2569–2590.

[32]   J. Powell and G. Maise, "Startram: The magnetic launch path to very low cost, very high volume launch to space," in *2008 14th Symposium on Electromagnetic Launch Technology*, 2008, pp. 1–7.

Chapter 6
Future electrostatic technologies

Jeremy Smallwood[1]

Electrostatics is often thought of as more of a scientific curiosity or 'old technology', for many people conjuring memories of rubbing things with cat fur and dramatic demonstrations of a Van der Graaf generator in school science lessons.

Electrostatics is possibly best known as a nuisance or hazard and for the field of static control and hazards in industries as diverse as electronics manufacture, fire and explosion hazards in handling flammable liquids and dusts, and of course in energetic materials (explosives and pyrotechnics). There is also the ubiquitous issue of electrostatic shocks to personnel in daily life [1]. Few of us have not from time to time experienced shocks at home, in the workplace, retail or other environments, or getting out of the car – the list of situations goes on. A whole industry has grown since the 1980s around static control, particularly for the electronics industry. One can now buy a wide range of footwear, flooring, workshop furniture, tools, personal protective equipment (PPE) and clothing, and materials designed for electrostatic discharge (ESD) control. Many companies set up ESD protected areas for electronic components and assembly manufacture or handling or they establish static control for the avoidance of fires or explosions. Offices, hospitality, retail and other establishments select floor and furnishing materials and other items to reduce the experience of shocks to workers or the public or to address other risks. There are International Standards and guidance documents to help the user address electrostatic issues in industrial processes. Despite this, incidents still from time to time make the news [2].

A different approach is taken in the prevention of ESD interference with working electronic equipment and systems. In this case, the ESD source is often the charged person in their domestic environment or workplace, who can become charged to several kilovolts in a normal uncontrolled environment. They might start to feel shocks if they discharge with a body voltage above about 2 kV [3]. (A person is relatively insensitive to ESD – for comparison, some types of electronic components can be damaged by a discharge from a person charged to 100 V or less.) Electronic systems placed on the market for use in these areas are required by EMC Directives in Europe to be designed to be relatively immune to ESD from

[1]Electrostatic Solutions Ltd., Southampton, UK

a person charged to up to 8 kV. They may be required to react predictably, if they do react. This is often validated and confirmed by ESD immunity tests.

There are some well-established technologies that use electrostatics for very practical applications. Among the best known of these is electrostatic spraying, which has found widespread application in painting and in agricultural applications of chemical treatments. Flocking (deposition of a layer of short fibres to form a carpet-like material) is another similar application. Electrostatic separation technologies have been used for many years in ore refinement, and more recently in waste recycling. Electrostatic separators can be used to separate materials based on their electrostatic and physical properties and can be used for tasks that are otherwise difficult to achieve.

This chapter focuses upon one application area of electrostatics – healthcare – that currently seems to flaunt missed opportunities and be ripe for reaping real benefits for a modest investment. Even though considerable work has been done, potential benefits have been demonstrated, and relatively low-cost and simple suitable solutions may already be available, this knowledge has not yet engendered adoption into standard practice. However, it can be glimpsed like a glow upon the electrostatic horizon.

6.1   Static electricity and its effects

6.1.1   How static electricity works

Science tells us that matter is made of atoms consisting of positive and negative electrical charges. In a neutral material, these are present in equal numbers of opposite polarity charges. When present in balance like this, no static electrical effects are observed. Yet even a relatively modest imbalance can conjure intense electrostatic fields and voltages.

Charges are separated when any two different materials come into intimate contact. Some charges move from one material to the other – this is known as triboelectrification [4,5]. If the contacting materials are then separated, one will have a net positive charge, and the other an equal negative charge.

If the material on which the excess charge resides allows the charge to move around (i.e. is a conductor with low or intermediate electrical resistance), the charge excess will tend to dissipate due to mutual repulsion of like charges and attraction to opposite polarity charges. If the material on which the charge resides does not allow the charge to easily flow away from the point of contact (i.e. is a good insulator with high electrical resistance), the charge can lurk for considerable time as static electricity. Conductors can also retain their charge if the charge dissipation path is blocked by an effective insulator. Repeated contact or rubbing against other materials can lead to the accumulation of charge if the charge is separated faster than it can be dissipated.

A local buildup of charge produces an increase in electrostatic potential and an electrostatic field that can affect the surroundings. The number of excess charges involved can be surprisingly small – the maximum charge density on a planar surface in air, giving the breakdown field strength of air of 3 MV/m, is just 2.64×10^{-5} C/m^2. This represents only a few charges per million atoms on the surface [4].

An electrostatic field can give various observable effects. Charged particles with a like polarity are repelled, and those with opposite polarity are attracted to one another due to the Coulomb force. It is not just charged items that are acted upon by force – neutral materials can also be attracted or repelled in a non-uniform field by dielectrophoretic force.

If the field strength is high enough, the breakdown of the insulating properties of the encompassing gas (air) can generate ESD. ESD between two conductors (a spark) or between a conductor and an insulating surface (brush discharge or propagating brush discharge) can be energetic enough to ignite some flammable materials such as solvent vapours and gases. Some sparks and propagating brush discharges can also be energetic enough to ignite some dust clouds [6]. Other types of ESD are possible in different situations. The electrostatic field is intensified by a small radius of curvature for the charged surface. A sharp edge or point in an electrostatic field can intensify the field to the point that charge is sprayed into the surrounding gas in a corona discharge. This effect can be used in ionizers to make local ion clouds, to charge neutral particles or to neutralize charged items.

If there are potential differences between nearby surfaces, then an electrostatic field will exist between them. The field can be represented by field lines (Figure 6.1). Each of these can be thought of as beginning and ending on a charge.

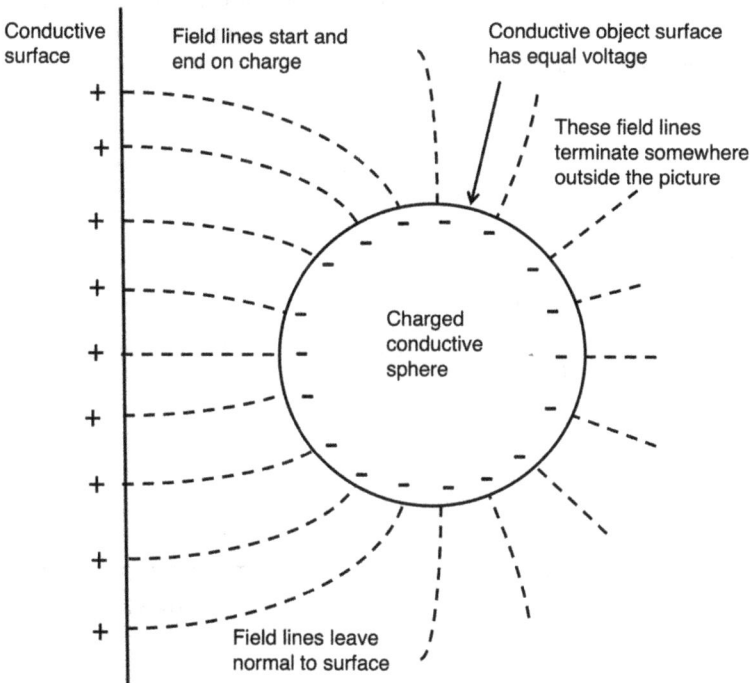

Figure 6.1   Field between a charged sphere and planar surface

A high charge density is then associated with high electric field strength. For a conducting surface, the field line exists normal to the surface, and in the static state, there will be no potential (voltage) difference across the surface (if there were, a current would flow until the voltage gradient on the conductor was returned to zero). In contrast, the voltage at any point on the surface of an insulator can be different from adjacent points, as current cannot easily flow across or through an insulator, which has very high electrical resistance.

6.1.2   Triboelectrification, grounding and earthing

There are few people nowadays who have not experienced electrostatic shocks in their everyday environments. These shocks are felt if the person's body exceeds around 2000 V, and they then discharge to a metal or other conductive item such as a door handle, filing cabinet or car door. Carts, trolleys and other mobile equipment can become charged as they are pushed around, and vehicles charge as they are driven around on some road surfaces. It is useful to understand why this phenomenon is so prevalent today.

The root of the issue is of course triboelectrification – any two types of materials contacting one another leads to separation of charge in the atoms on the surfaces of the materials. The charge generated on both materials will try to dissipate and eventually recombine to be neutralized. If it cannot dissipate, charge builds up and can reach the point where it creates electrostatic effects, or ESD occurs.

The simple circuit of Figure 6.2 helps to explain the principles of how electrostatic charging and ESD can occur in many situations. A more detailed model reflecting real-life situations might of course be considerably more complex.

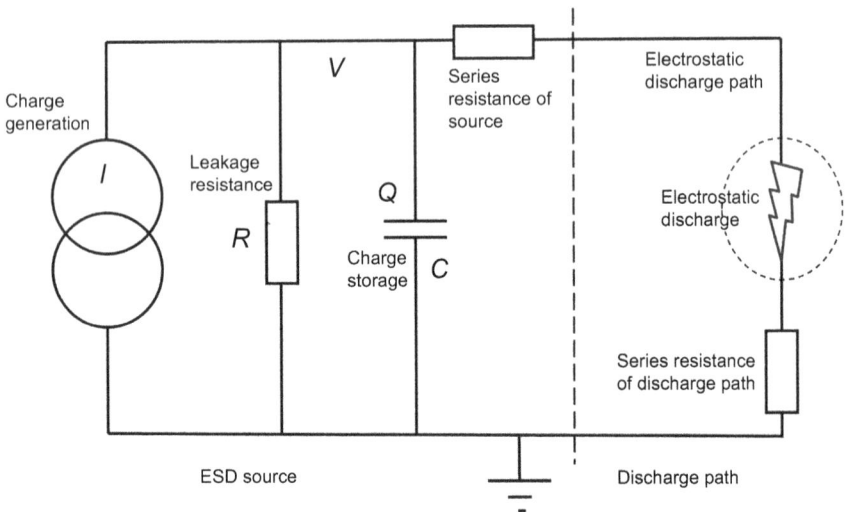

Figure 6.2   Simple model of electrostatic charging and ESD generation

Triboelectric charge generation effectively represents a current flow generated, for example, between tires and paving as mobile equipment rolls or between shoe soles and floor as a person walks. The charge dissipates, usually to earth, via a material that has a resistance R. Charge is also stored, represented here by a capacitance C between the surfaces. The stored charge on the capacitance has a corresponding voltage V. Current flow I through the resistance follows Ohm's law. In a steady state with V neither increasing nor decreasing

$$V = IR$$

So, neglecting, for the moment, the effect of capacitance C, the voltage V generated for a given charging current I is proportional to the resistance R of the dissipation path, often referred to as the 'resistance to ground' or 'leakage resistance'. If the charging current is 10 nA, and the resistance to ground is 100 MΩ, only 1 V will be measured. This small voltage is unlikely to be noticed in a practical situation. If, however, R = 100 GΩ, V would be 1000 V, which might well be perceptible. If R is increased to 1 TΩ, the voltage produced of 10 kV is likely to be noticed and might give rise to ESD, if the breakdown voltage of the discharge path is exceeded. These days, we frequently use materials (e.g. polymers) that present resistance to ground well over 1 TΩ for flooring, shoe soles and wheel tires of mobile equipment. It should be no surprise, then, that we charge up as we walk around, and likewise mobile equipment as it is pushed around. The case of cars and other road vehicles is different in that the resistance of their tires is perhaps surprisingly low in most cases around the MΩ region. Concrete, a commonly used floor material for some roads, often has resistance to ground of around tens to hundreds of MΩ. So, the voltage generated by a slow-moving car on a concrete road surface is likely to be too low to cause electrostatic shocks. Conversely, an asphalt surface might have resistance near 1 TΩ and give a much higher charging voltage – this has been the cause of shocks to personnel in car park facilities.

Considering the effects of charge storage on the capacitance C, we can see that if charge generation ceases abruptly, the stored charge on C leaks away through R with a voltage decay time constant RC. The voltage is reduced to 37% (i.e. $1/e$) of its initial value in a time RC, and about 5% in $3RC$. This shows us that if resistance increases by a factor of 1000, the voltage remains evident for 1000 times longer. An item having a capacitance of 10 pF, with a leakage resistance of 100 GΩ, shows voltage decay to 5% in 3 s. However, for a large fuel tanker vehicle having a capacitance of 1000 pF standing on asphalt with resistance around 1 TΩ, the decay time to 5% would be about 1000 s! The vehicle might remain significantly charged when the driver subsequently connects an earth cable or hose, possibly in the presence of a flammable atmosphere. Yet standing on a concrete floor with a resistance of 10 MΩ, the decay time would be only 0.01 s. The vehicle would lose its charge before the driver could exit the cab. This is why vehicle stands in fuelling station forecourts are normally concrete or some other partially conducting material [7]. The capacitance of conductors varies with size and other parameters. The approximate capacitance range of various common items is given in Table 6.1.

Table 6.1   Approximate capacitance range of various items

Item	Approximate capacitance range
Small items – nails, screws, cans	1–10 pF
Small containers (e.g. bucket)	10–100 pF
Human body	80–300 pF
Hospital beds	200 pF
Vehicles	800–1200 pF

The energy in a discharge is an important parameter in hazard evaluation. For example, in ignition risk evaluation, the likely energy in a spark discharge between two conductors can be compared with the minimum ignition energy required for ignition of flammable material. The energy E stored on a conductor of capacitance C at voltage V is given by

$$E = C V^2 / 2$$

Unfortunately, the energy of a brush discharge from an insulator surface is not easily calculated [6].

In static electricity control, the words 'earth' and 'ground' are normally considered synonymous, as are 'earthing' and 'grounding'. To prevent static charge buildup on conductors, a conduction path to ground (earth) is provided. The discussion has shown that the path might be via a dissipative material or flooring, although a wire can also be used. The maximum resistance to ground needed for an effective ground depends on the application, and particularly the charging current expected (or sometimes the charge decay time required). In IEC 60079-32-1 [6], for grounding personnel in the presence of a flammable atmosphere, the resistance from the body through footwear and flooring to the ground is specified as less than 100 MΩ. This requires that the user wear suitable static dissipative or conductive footwear as well as stand on a static dissipative or conductive floor. A similar arrangement is required for work in the electronics industry ESD protected areas, although the upper limit is 1 GΩ and a walk test of the footwear and flooring system must be done to ensure body voltage remains below 100 V. A wrist strap giving maximum resistance from the body to the ground of 35 MΩ is used for seated personnel [8].

The polarity of charging that occurs when two items rub together or make contact depends on the contacting materials. These can be ranked in a triboelectric series [4,5]. In theory, if a material is rubbed against one higher up in the series (Table 6.2), the higher one will charge positive and the lower one negative. The strength of charging tends to increase with combinations relatively more widely separated in the series. So, if glass is rubbed against Polytetrafluoroethylene (PTFE), the PTFE would be expected to charge negatively and the glass positively. If glass were rubbed against polyamide (nylon), the nylon would be expected to charge negatively, but if rubbed against polyethylene terephthalate (PET), polyamide would be expected to charge positively. These series can be somewhat

Table 6.2 A simple triboelectric series

Mica	
Glass	More positive
Polyamide (nylon)	
Wool	▲
Polymethyl methacrylate (PMMA)	
Silk	
Paper	
Cotton	
Steel	
Wood	
Rubber	
Acetate rayon	
Polyethylene and polypropylene	
PET	
PVC	
Polyurethane	▼
PTFE	More negative

variable and unreliable as there are many other factors involved than the material type. Also, although it might be surmised that a material rubbed against another sample of itself would not charge up, experience shows this is not the case – two samples of polyethylene rubbed against each other, or polyethylene film on a reel, can become strongly charged. Given that these materials are common in use, it is not surprising that modern environments are full of materials charged to different levels and polarities and surrounded by their resultant electrostatic fields.

Atmospheric humidity is another factor which is well known to have a considerable effect on electrostatic phenomena. The reason for this is that moisture is an excellent conductor of static electricity. At higher relative humidity above about 30% RH, a thin moisture layer forms on most materials. This tends to provide a path for the electrostatic charge to be dissipated to earth. Electrostatic charge buildup tends to be less evident as the humidity level increases. Below about 30% RH, there may not be sufficient moisture present to form a continuous surface moisture layer on materials. Static electricity issues tend to be greater in these low-humidity conditions.

6.1.3 Induction

Induction occurs when an ungrounded conductor is within the electrostatic field of a charged object. The field induces a voltage on the conductor depending on the amount of field coupling to the conductor. A measurement instrument such as an electrostatic field meter can detect the voltage (Figure 6.3) although the net charge on the metal plate is zero. Some positive and negative charges are merely separated by the field.

Figure 6.4 shows a simple electronic model of the situation as a capacitive divider.

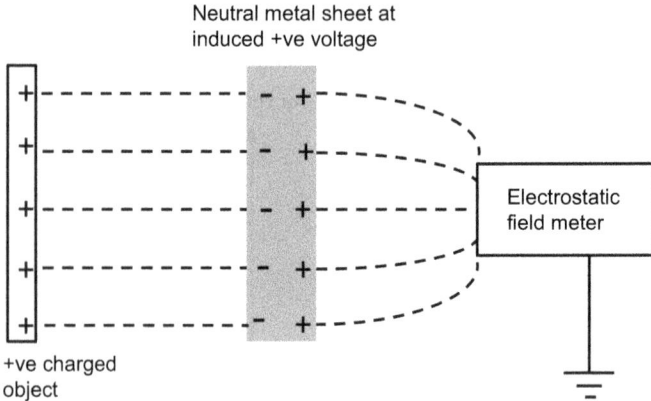

Figure 6.3   The field meter sees a high voltage induced on this metal plate by the field from a nearby charged object

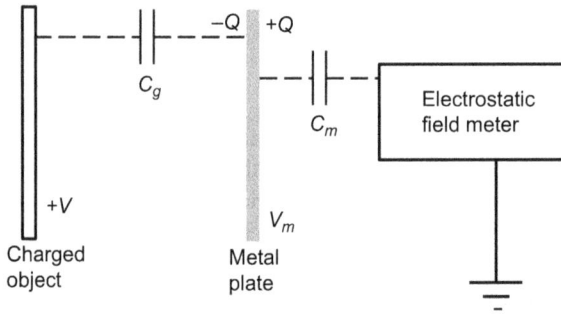

Figure 6.4   A simple capacitor divider model showing how high voltage is induced on the metal plate of Figure 6.3

When the metal plate is grounded, the plate voltage goes to zero, and the plate is left with a net negative charge despite the zero voltage (Figure 6.5).

If the ground wire connection is then removed so the plate is left isolated, it is left with the net negative charge while still being at zero voltage. If the charged object that produced the original electrostatic field is then also removed, the plate is left with both a net negative charge and voltage (Figure 6.6).

This often appears paradoxical at first sight but is a common phenomenon in real situations where ungrounded conductors are present. For example, in one case this author investigated, high voltages were induced on operators' bodies while unrolling polyethylene from a large reel of packaging film. The high voltage caused breakdown and ESD between their feet and the concrete floor, giving them shocks to the feet. Once this had happened, they were themselves charged and could receive a second shock, this time to the hand on touching another person or a metal rack or equipment.

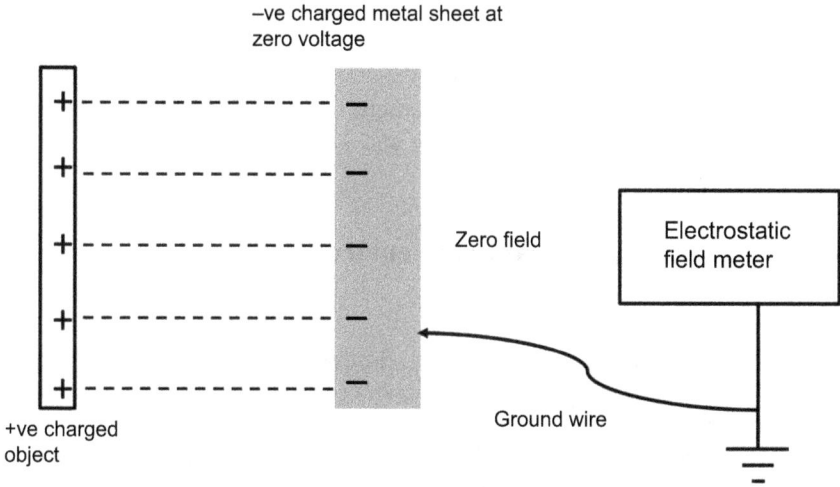

Figure 6.5 When the metal plate of Figure 6.3 is grounded, the positive charge leaves and it goes to 0 V while becoming negatively charged

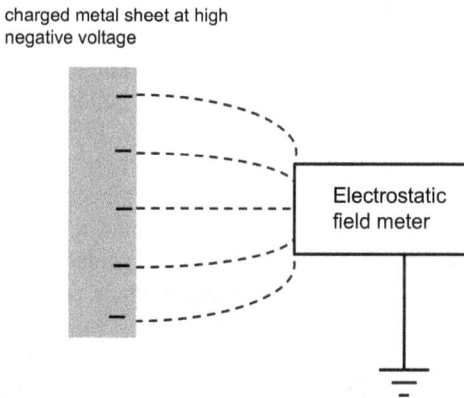

Figure 6.6 When the ground wire of Figure 6.5 is removed and the charged object taken away, the field meter sees the metal plate is charged to some negative voltage

6.1.4 Movement of particles in an electrostatic field

A charged particle entering an electrostatic field E experiences a force F depending on the field strength and direction and the particle charge q.

$$F = qE$$

In the absence of other influences, the force would cause the particle to follow the field line towards one of the terminating surfaces, in a sense according to its polarity and the field direction.

Dielectrophoretic particle action is rather more complex and depends on the divergence of the electrostatic field lines as well as the relative permittivity of the particle and the surrounding medium [4].

The phenomenon of attraction of particles to a surface due to electrostatic fields is known as electrostatic attraction and is important in many applications where cleanliness is important.

6.2　Electrostatics in healthcare environments

6.2.1　Background

One of the key drivers for concern about electrostatics in hospitals has been the accidental ignition of flammable anaesthetics in operating theatres, which has caused explosions. This has led to the adoption of electrostatic controls including the installation of conductive floors [9–11]. The most recent guidance document [9] gave a thorough review and explanation of electrostatic issues. It stated that minimizing electrostatic charge in the hospital environment is important in areas where potentially flammable gas mixtures are present and where sensitive electronic equipment is in use. However, in its conclusions and recommendations, it stated that the NHS Procurement Directorate Working Party [12] had indicated that provision of antistatic facilities was no longer required in operating theatres where flammable anaesthetics would not be used, as there was a cost saving in discontinuing the practice. However, 'consideration should be given to minimizing the buildup of static electricity by careful choice of materials and environmental controls' [9]. It indicated 'a remote risk of damage to electronic equipment . . . ' by ESD and 'cost savings can be achieved by quantitative definition of the static hazard (if it exists) in any operational environment and then implementing only the necessary countermeasures'.

It is perhaps unsurprising that since then, UK guidance on avoidance of electrostatic charge buildup in the healthcare environment seems to have almost disappeared [13]. References to static electricity seem to indicate only that it should be considered where flammable anaesthetics are to be used. Nevertheless, electrostatic shocks to personnel and ESD interference with equipment remain an important issue [14,15]. Ignition of materials other than anaesthetics also continues to occur, with ignition of hand sanitizing gel being reported during the recent COVID-19 pandemic. COVID-19 has also stimulated interest in medical issues generally, and notably technologies such as electrostatic spraying of sanitizing solutions [16].

In 2000, hospital-acquired infections were estimated to cost the NHS in England £1 billion per annum [17]. The contribution that electrostatics technology could make in reducing this seems to have been largely neglected in subsequent years.

The IEC have recently issued general guidelines in the form of the IEC 61340-6-1 standard [18] for control of static electricity in healthcare facilities, led by Scandinavian countries which experience high levels of static electricity in the winter months exacerbated by low humidity conditions.

Jamieson *et al.* [19] reviewed several benefits achievable for healthcare and other sectors in taking 'electromagnetic hygiene' into account. One of their conclusions was that the development of novel (often low cost) technologies and techniques could dramatically reduce the socio-economic burden of hospital-acquired infection whilst increasing well-being and work efficiency of society in general. Their paper considered several aspects including electromagnetic and electrostatic fields, triboelectric charging, charge and particle deposition, grounding regimes, bipolar ionization and humidity levels. Some of these aspects will be further discussed here.

6.2.2 Particle deposition and infection control

Hospital-acquired infections pose significant challenges due to compromised immunity of some patients and invasive medical procedures. They lead to increased mortality rates and healthcare costs [20].

Indoor air particle sizes vary in diameter from about 0.01 upwards to more than 100 μm for airborne soil, allergens and pollens [21]. Viruses can be in the <0.01–0.31 μm range, and bacteria lie in the range of 0.05–10 μm. Submicron contaminants are associated with an increase in cardiovascular problems, respiratory diseases such as asthma and chronic obstructive pulmonary disease and may increase the risk of infection from pathogens [19]. The charge of biological particles can be much greater than that of non-biological particles, although this also depends on the means of their aerosolization [22].

Those who remember cathode ray tube (CRT) TV displays will recall how quickly and easily the screen became dusty in use. The reason, of course, was that the high electrostatic field associated with the CRT attracted dust particles from the air, which were deposited and adhered to the screen.

Deposition of particles to an electrostatically charged chamber wall has been found to be dominated by electrostatic effects for particle size between 0.05 and 1 μm [23]. Wall loss rates of singly charged 0.1 μm diameter particles were a factor of 100 greater than neutral particles of the same size. Particles of either polarity were deposited at the same rate.

In nature, a vertical potential gradient (VPG, electrostatic field) around 80–150 V/m exists during fair weather conditions, with the earth at negative polarity [19]. The gradient can be increased in elevated locations such as mountains. The VPG can of course become much steeper in thunderstorms and can reverse direction under some conditions. Most buildings are made of sufficiently conductive materials (e.g. steel framed and concrete structures) to constitute an effective Faraday cage, with little or no VPG penetration. The presence of electrostatic field sources such as charged materials or some types of electronic equipment can modify a building's internal electrostatic fields.

Jamieson *et al.* [19] reported work from several authors that indicate improved immune system response in animals with a field strength of just 40 V/m. They suggested that similar effects might occur for humans, citing some evidence supporting this, including reduced infection rates and improved recovery rates in hospital areas with ceiling artificial VPG systems installed. Airborne microbe concentrations were reported to be reduced by about 95%. The use of VPGs in a situation with no ventilation reduced airborne particles of diameter less than 250 μm by about 60%. The combined effect of VPGs with natural ventilation was more efficient than the use of natural ventilation alone.

In modern times, many items and fabrics are made of highly insulating polymer materials. It is not unusual for these to become highly charged and generate strong electrostatic fields. Some types of equipment, such as CRT displays or televisions, have historically been sources of strong fields. Equipment structures and housings, furnishings and drapes are often made of polymer materials.

High electrostatic fields can lead to surface deposition of particles, local reduction of air ions and changes to the local concentration of ultrafine particles. There is evidence that significant benefits might be obtained simply by using low charging, static dissipative or conducting materials rather than highly insulating materials, thus eliminating or reducing electrostatic charging, associated fields and virus and bacteria deposition on human interactive surfaces. Figure 6.1 shows that when a voltage difference exists between two items, an electrostatic field exists between them. Particles will migrate to both surfaces in response to the forces induced by the field and can be deposited on either surface according to their polarity.

Allen *et al.* [24–27] found that electrostatically charged plastic aprons worn by nursing staff can increase microbial deposition in patients. Polyethylene aprons attracted 83% more bacteria to their surface compared to 17% by aluminium foil aprons [27]. She reported that 95% of transplant patients develop potentially life-threatening infections, and hospital infections were the eventual cause of death in 10–15% of these patients. The plastic materials widely in use in hospitals are prone to charging easily and considerably and can be very efficient at collecting airborne micro-organisms. Deposition is dependent on electrostatic fields and increases with surface voltage. A nurse wearing a charged plastic apron will set up an electrostatic field between the nurse and the patient. This can be expected to lead to the deposition of micro-organisms onto the patient.

Plastic materials are in widespread use in hospitals. More than 5 kV is readily induced on an apron by tearing it off a roll. Plastic items in common use were investigated and found to charge to voltages ranging from a few hundred volts to kilovolts by rubbing that mimicked normal handling. Voltage decay times were found to stretch from a few minutes to several hours.

Although similar research has not, to my knowledge, been carried out on other items of PPE (e.g. gowns, scrub wear, respirators, face shields, visors, etc.), it is reasonable to suspect that electrostatic charge generated on other items of PPE worn by staff might create similar infection hazards. Furthermore, it is also reasonable to suspect that the use of face shields and visors made of insulating plastic sheets, which can retain electrostatic charge over protracted periods, might attract airborne pathogens towards the wearer's airways, which is clearly undesirable.

Allen *et al.* [25] found that electrostatic charging may contribute to hospital-acquired infections in a bone marrow ward. In this case, the patients were nursed in an environment with filtered air and rigorous infection control. They suffered from immunodeficiency due to high-dose chemotherapy and radiotherapy. Skin breach sites act as portals for infection. Skin scales shed by hospital staff are a major source of micro-organisms. The researchers explored the effect on surfaces of static charge by suspending conducting plates charged to 4 kV in the ward. Up to a 20 times increase in bacterial deposition was observed – equally for positive and negative charging polarity.

Static charge might lead to infection in several ways. Tearing open sterile plastic packaging could charge the contents and the packaging and attract airborne bacteria to them. Bacteria deposited on aprons could be transferred by contact. A nurse wearing a charged apron will set up an electric field between the nurse and a nearby patient. Airborne bacteria can carry a charge and are susceptible to movement in electric fields. The bacteria in the air could be attracted to a charged up patient.

Other authors have pointed to a possible role of electrostatic charge in mediating hospital infections. Todd *et al.* [28] measured the electrostatic charge on six plastic-giving set needles. Values ranged from $+0.45 \times 10^{-8}$ to 1.9×10^{-8} C corresponding to 1–5 kV for an object of typical dimensions around 5 cm.

Becker *et al.* [29] found that during open surgery, the act of pointing to a visual display unit (VDU) screen for teaching purposes significantly increased the deposition of bacteria on a surgeon's gloves. This would be expected due to the concentration of the electrostatic field at the fingertip. Treatment of latex drains, silicon drains and plastic coverings with antistatic solution to reduce charging resulted in the deposition of fewer bacteria than were found on items left untreated.

Plastic film with a conducting coating that prevents buildup of charge is manufactured for various uses. Nowadays, a plethora of materials are available that have low charging or static dissipative properties, often developed for ESD control in electronics manufacture. Nurses' aprons and other items could be made of some of these materials. Decreasing the risk of potential infection may have major effects on the morbidity and mortality of transplant patients and might lead to considerable cost savings as well.

In some cases, the benefits of electrostatic control could be achieved by processing insulators with an 'antistatic' treatment. Cozantis *et al.* [30] used an alcoholic benzalkonium solution to treat stopcocks, latex drains, silicon drains and plastic coverings. They found that static electricity did not accumulate when the material resistance was less than 10 GΩ following treatment. Their results suggested that airborne bacterial contamination was significantly reduced for antistatic-treated insulating items.

Despite the evidence, there seems to have been little or no response from the healthcare sector so far. Procurement specifications in the NHS do not include antistatic requirements for personal protective equipment (PPE). The electrostatic requirements that do exist for any other hospital equipment relate only to risks of igniting flammable anaesthetic gases. The European Standard for protective clothing against infective agents EN 14126 [31] does not include any electrostatic requirements. However, there is an informative Annex that discusses ESD damage to equipment and ignition of flammable aesthetics and vapours [32]. Likewise, the European Standards for surgical clothing and drapes EN13795-1 [33] and EN 13795-2 [34] are lacking in any electrostatic requirements.

6.2.3   Electrostatic charging of people and mobile equipment

Healthcare facilities provide many opportunities for people and mobile equipment to become charged with static electricity, leading to shocks experienced by personnel, risk of ignition of flammable materials and damage, upset or data loss to electronic equipment. One type of shoe had to be banned and featured in the national news in 2007, being reported to 'generate static electricity ... strong enough to knock out respirators and machines in operating theatres' [35].

Shocks to personnel can happen in two main ways. First, a person can charge up, e.g. while walking or while rising from sitting [13,15]. They then get a shock when they discharge to a conductor if their body voltage grew greater than about 2 kV. Second, a conductor such as a large metal bed, trolley, wheelchair or other equipment can become charged and discharge to the person. A bed can become highly charged simply by stripping off bedding material [13]. A nurse or patient can become charged in the same process. In one case, Holdstock found, when a flame-retardant (FR) polyester counterpane was pulled from a modacrylic blanket, the puller's body voltage rose to over 60 kV. This would be sufficient to ignite flammable atmospheres, deliver a painful shock to the person or damage medical equipment. It was observed that the use of fabric conditioners usually resulted in reduced charging, although none of the conditioners used reduced charging below hazard voltage thresholds. Conditioners had an adverse effect on the charging of a combination of FR polyester counterpane and FR polyester blanket.

Rolling items can become charged during pushing from one position to another, and a person holding the item can become charged at the same time. Issues of this nature are exacerbated in a low-humidity atmosphere.

High voltages on mobile equipment and personnel can be avoided if they can be grounded (earthed). For personnel, this can be done via ESD control footwear and flooring if these are provided. Nowadays, a wide range of suitable footwear is available, such as 'conductive' or 'antistatic' footwear accredited to ISO 20345 [36]. In areas where personnel wear their own choice of shoe, the grounding of personnel is less straightforward. Nevertheless, the provision of a static dissipative floor can help reduce body voltage buildup due to walking across the floor.

For mobile equipment, grounding can be achieved if it is so designed that the metal and other conducting parts are connected via conducting wheels to an ESD control floor.

6.2.4   Malfunction of electronics equipment

In the latter part of the twentieth century, ESD interference with electronic systems became a familiar hazard. As a researcher in the early 1990s, the author frequently sat down at his computer workstation, picked up the computer mouse and promptly crashed the computer network, to the irritation of his colleagues! The EMC Directive [37] brought significant improvements to this situation, requiring electronic products placed on the market to have a level of immunity to ESD [38,41]. This is tested using the IEC 61000-4-2 ESD test [39], which effectively simulates

ESD produced by a person charged to different voltage levels. The ESD generator consists of a simple circuit in which a 150 pF capacitor discharges through a 330 Ω resistor, typically switched by a vacuum relay. The series resistance models the resistance of the human body. Other circuit elements are typically present, giving an output waveform that is unidirectional with a fast rise time in the nanosecond range and with an initial high current peak. The model is sometimes known as the 'human-metal model' [40]. Test voltages up to ±8 kV are typically used in testing the immunity of medical equipment [41].

Unfortunately, experience shows this has not prevented electrostatically induced malfunctions of medical devices [14,40]. In recent times, the medical industry has developed and used ever greater numbers of medical devices with improved functionality. ESD in healthcare situations continues to upset equipment or cause damage or data loss or corruption. Kohani *et al.* [14] data mined the US Food and Drug Administration (FDA) and Manufacturer And User Facility Device Experience (MAUDE) databases between 2006 and 2016 to determine the number and types of malfunctions and their impact on patient health. The consensus is that these malfunctions are underreported and not well understood. Their conclusions included that manufacturers of medical devices do not implement adequate ESD prevention, and the reports of malfunctions, recalled devices and patient complications indicate that ESD immunity standards for medical devices might not be adequate. Analysis of the severity of each failure mode should include possible patient complications. Both recoverable malfunctions (such as data corruption) and irrecoverable ones (such as device shutdown) can result in patient injury or death. Superficially non-critical malfunctions such as time and date resets, intermittent communications drops and changes in sensor readings could also cause patient complications. These might not damage the devices, but cases of consequential patient injury and death have been reported. The number of ESD malfunctions during cold, dry months was nearly six times higher than in more humid summer months. This could compel healthcare facilities to maintain a higher humidity level: at least 30% RH.

Manufacturers could often anticipate reported device failures by conducting ESD tests at higher than the maximum voltage specified in the standards. Tests should realistically simulate charging and discharging experienced in daily activities, such as touching a keypad or screen, rolling out of bed and plugging connectors into a device.

Viheriaekoski *et al.* [40] help to explain why current ESD test sources might inadequately represent the risks seen in healthcare facilities. While current equipment ESD tests are based on a human body ESD model, in healthcare facilities common ESD sources include patient beds, carts, trolleys and intravenous stands. These were often ungrounded metal objects. The actual ESD in this case can be a metal–metal discharge with little intervening resistance to limit the current flow or absorb ESD energy. Higher discharge power and energy, and greater radiated electromagnetic emissions tend to result. An item such as a bed frame could have higher capacitance than the standard ESD model, and higher voltages up to 30 kV were estimated. A frame was charged by being wheeled around. A discharge from the standard ESD model (150 pF) charged to 8 kV represents 4.8 mJ of stored energy, much of which is absorbed by ESD current flow through the 330 Ω series resistance in the standard model discharge circuit. Conversely, a metal patient bed with an estimated capacitance of 200 pF at the highest observed voltage of 30 kV represented 90 mJ of stored energy – with little series resistance in a metal–metal discharge much of this energy would potentially be delivered to the ESD victim device or person.

6.2.5   Electrostatic spraying of disinfectant

According to Robertson [42], less than 50% of environmental surfaces in patient care rooms are properly cleaned and disinfected. High-touch surfaces are identified as important for the transmission of pathogens. Electrostatic spraying is an example of an old technology that could be finding important new post-pandemic applications in healthcare [20]. Since the advent of the COVID-19 pandemic, users have been looking for ways of disinfecting large areas of facilities quickly, regularly and effectively [43], not just for healthcare facilities but also homes, businesses, schools and transport. Electrostatic spraying potentially offers advantages in that charged sprayed aerosol particles could be attracted to the target surface, with improved disinfectant deposition and reduced loss to the surroundings. The aerosol particles could potentially follow electrostatic field lines that wrap around into more hidden parts of the surface. As an established technology, it is not surprising that evaluation has often focused on spraying equipment from the existing market. The materials to be sprayed can include papers, ceramics, metals, glasses, plastics, fabric and wood. Otherwise, pathogens can survive on these from seconds to hours or even for days.

The two main types of the existing sprayer are those that are powered from an electricity outlet, and those that are battery powered [16]. AC outlet-powered units can include a compressor to give more reliable droplet size and charging and hence improved performance. They can also be grounded via the outlet connection. Grounding can be required for safety and for maintaining charge neutrality in the electrostatic spray apparatus. If an apparatus produces charged spray of predominantly one polarity, the charging unit will charge to the opposite polarity unless this charge is able to dissipate to ground. Two other influential factors in the effectiveness of electrostatic spraying are the particle average volume mean diameter (VMD) and the particle charge-to-mass ratio. Smaller droplets can achieve higher charge-to-mass ratios and are more likely to respond to electrostatic fields. A minimum charge-to-mass ratio of about 1.0 mC/kg is needed to achieve the wrap-around effect [43]. Droplet size depends on parameters such as applied voltage, air pressure, liquid flow rate, liquid properties and surface tension.

Electrostatic attraction of particles to the target means that less disinfectant could be used to cover a surface, although there needs to be sufficient for the surface to remain wet for long enough to be effective in destroying pathogens [43,44].

Evaluation of the technology should of course include safety, functionality and ease of use in the intended environment as well as proof of effectiveness in terms such as surface coverage, compatibility with surfaces treated and reduction in viable microorganisms after treatment. These points together with the safety of application and any PPE required depend on the disinfectant used as well as the spray system. There can also be regulatory hurdles to overcome before electrostatic spraying can become established practice.

Battery-powered units can of course confer convenient portability with no power cord but require regular charging. Nevertheless, they must be grounded via the operator or by some other means, and some attempts to achieve this can engender inconsistent grounding. If not grounded, they are likely to charge up during spraying to the opposite polarity to that of the spray and cause inconsistent results.

Wood *et al.* [44] evaluated a selection of sprayers and foggers from the commercial market with three disinfectants. Most devices gave droplet VMD in the range of 10–100 μm, with an average VMD of at least 40 μm. For electrostatic sprayers, the VMD tended to reduce with increasing distance, presumably as larger particles dropped out. Four out of the six units tested produced spray with a charge-to-mass ratio of at least 0.1 mC/kg. The units that gave the largest magnitude charge-to-mass ratio (-3.6 to -6.0 mC/kg) were AC powered and gave negatively charged spray. Four battery-powered units gave positive polarity with charge-to-mass ratio about an order of magnitude lower. They observed minimal wrap-around of the spray when used on an 8-in. (203-mm) diameter cylindrical target even with the highest droplet charge-to-mass ratio. The loss of the spray active ingredient to the surrounding air was minimal.

Chauhan *et al.* [43] evaluated an air-assisted electrostatic spray disinfection equipment using sodium hypochlorite on five surfaces – wood, glass, stainless steel, plastic and fabric. The equipment used a main-powered high-voltage generator with compressed air to produce the spray. Charge-to-mass ratio over 2.4 mC/kg was obtained with droplets around 30–60 μm, five to six times smaller than obtained with a conventional backpack sprayer. The particle size range and charge-to-mass ratio were deemed suitable for the effective killing of pathogens and obtaining wrap-around effect. The spray volume was about ten times lower than a conventional backpack sprayer. Material wastage was reduced by a factor of approximately 12.8.

One potential factor that does not seem to have been addressed by most investigators is the possibility that the target materials could be highly charged. An exception is Robertson [42] who stated that most surfaces are neutral or negatively charged, and so a positive charge should be placed on the spray particles. The justification for this assertion is not entirely clear, although, from the triboelectric series, many polymers are expected to charge negatively when rubbed by other materials. There are exceptions to this such as nylon, which can often charge positively. Clearly, an initially highly charged target material could repel like polarity droplets, which might lead to reduced deposition in some circumstances. Conversely, opposite polarity droplet deposition could be enhanced. For the most reliable coverage, neutral target materials that are uncharged could be ideal.

6.2.6   Ignition of flammable materials

Electrostatic precautions continue to be required in areas where flammable anesthetics might be used, or flammable atmosphere zones are present.

With the recent widespread use of alcohol-based hand sanitizer products during the COVID-19 pandemic, there have been reports of ignitions of the alcohol vapor by static electricity leading to burns [45,46,47]. Alcohols can form near optimum ignitable mixtures with air near hospital ambient temperatures. Ignitions are, however, surprisingly rare. This is likely to be because it requires the coincidence of the user generating sufficiently energetic ESD to ignite the mixture during the short time that sufficient vapor is present for ignition. Nevertheless, spillage or soaking of the sanitizer into fabrics could protract the risk. The presence of oxygen in some healthcare situations could considerably increase the likelihood of ignition, as oxygenated vapours are ignited by much lower energy ESD.

6.2.7   The IEC 61340-6-1 standard

The IEC 61340-6-1 standard [18] was developed and published in 2018, having been proposed by the Finnish National Committee. Finland, in common with other Scandinavian countries, experiences very low atmospheric humidity in winter and often experiences consequentially high levels of static electricity. The Finnish proposal was based on reports of medical equipment failure due to ESD, accounts of ignition of alcohol-based hand sanitizer and concerns over contamination due to electrostatic attraction (ESA) [32]. The standard considers ESD damage or disruption to medical and data processing equipment, damage to electronic equipment during service and maintenance, contamination due to ESA, ignition of flammable substances and shocks to personnel.

Facility locations are classified according to IEC 60364-7-710 [48] as

• Unclassified – e.g. waiting rooms, office areas, corridors, etc.
• Group G0 – Electrostatic control recommended. Consulting rooms, inpatient wards.
• Group G1 – Electrostatic control recommended. Endoscopic examination rooms, Electroencephalogram, electrocardiogram, electrohysterogram, computer tomography, special care baby units, etc.
• Group G2 – Electrostatic control required. Operating theatres, preparation and recovery rooms, coronary care units, intensive care units, etc.

The use of various control methods in the standard is summarised in Table 6.3.

IEC 61340-6-1 classifies materials for electrostatic control based on their electrical resistance, charge decay or triboelectric characteristics. Materials are classified as 'conductive' if they have surface or volume resistance less than 10 $k\Omega$ measured in accordance with IEC 61340-2-3. 'Dissipative' materials have resistance between 10 $k\Omega$ and 100 $G\Omega$ and a charge decay time of less than 2 s. (The charge decay time is measured as the time taken for a surface voltage between 200 and 100 V to decay to less than 100 V.) Alternatively, on charging by corona discharge, the surface remains below 190 V measured in accordance

Table 6.3 Summary of electrostatic control methods given in IEC 61340-6-1

Location	Control method			
	Ground personnel via footwear and flooring	**Ground other conductors via floor or direct connection**	**Use of grounded conductive or dissipative materials**	**Use of low-charging materials**
Unclassified	Not mandatory	Not mandatory	Not mandatory	Recommended
G0	Recommended	Recommended	Recommended	Recommended
G1	Recommended	Recommended	Recommended	Recommended
G2	Required	Required	Recommended	Recommended

with IEC 61340-2-1 [49]. A material is classified as 'insulating' if the resistance is greater than 100 GΩ.

A floor required for grounding personnel is required to have resistance to ground (earth) less than 1 GΩ or less than 1 MΩ in locations where flammable anesthetics and hyperbaric oxygen systems are used, and where high charge-generating mechanisms are expected. In locations where static control is required but it is not possible to ground personnel, flooring is specified to produce less than 2 kV on a test subject in a walking test in accordance with IEC 61340-4-5 [50].

Where footwear is required for grounding personnel, the footwear is specified to have resistance less than 1 GΩ, or 1 MΩ in locations where flammable anesthetics and hyperbaric oxygen systems are used.

In locations where electrostatic control is required, mobile equipment such as patient beds, stretchers, intravenous stands, trolleys and carts and chairs are required to be connected to the ground with resistance to ground [51] less than or equal to 1 GΩ. This can be achieved via at least two dissipative or conductive wheels and an ESD control floor.

Textiles for static control are specified to be made from conductive, dissipative or low-charging materials. (Low-charging materials are classified as those not exceeding a voltage of 1 kV or a charge density of 2 $\mu C/m^2$ in a friction charging test.) They are required to have one or more of the following characteristics:

• Surface resistance less than 100 GΩ.
• Charge decay time from the initial value between 200 and 1000 V down to 100 V in 2 s or less, or initial corona charging voltage <190 V (IEC 61340-2-1).
• Friction charged surface voltage not exceeding 1000 V (ISO 18080-2 [52] or ISO 18081-4 [53]).
• Surface charge density not exceeding 2 $\mu C/m$ (ISO 18080-3 [54]).

Items such as aprons and packaging are also specified to be made from conductive, dissipative or low-charging materials to reduce the deposition of microorganisms and production of electrostatic fields on nearby people and surfaces.

6.3   Future benefits

In the mid-nineteenth century, infection control procedures we now consider routine and essential such as washing hands were yet to be adopted [55]. The theories that disease could be spread by hand contamination and faecal oral transmission were initially rejected. Thankfully the efforts of pioneers brought now widely recognized simple countermeasures into common practice.

The importance of electrostatics in the healthcare environment unfortunately seems currently similarly widely dismissed or underestimated. Electrostatic control measures seem only to be considered where the risk of ignition of flammable materials, e.g. of flammable anesthetics, is possible. Elsewhere, it seems to be considered a needless expense.

Electrostatic control can be achieved by choice of static dissipative or conductive materials used in equipment, furnishings and fabrics. Fifty years ago, these materials were certainly less available than they are today. The fledgling ESD control industry serving electronics manufacture had to make do with limited ESD control products and materials. Yet in consequence, there is now a plethora of ESD control products on the market and in routine use in the electronics industry. Many of these have become widely used for ESD control in other process industries as well. It is not just industry that uses static control measures. There are standards that specify static electrical performance of floor coverings for general-purpose uses on office, retail or domestic premises [56].

The healthcare world depends on standards and best practice guidance for the specification of materials and PPE, but these rarely specify general static control. In a conversation with my local pharmacist a few years ago, the pharmacist asked why he and his patients suffered static shocks daily in his consulting room. A quick view of the room immediately showed the answer: both the floor and chairs were made of plastics, which could be expected to lead to electrostatic charge buildup. The pharmacist explained that he had to use these materials and items which were specified according to approved best practice – no doubt part of this approval was consideration of easy cleaning and hygiene, but evidently electrostatic suitability was not a criterion.

It seems clear that greater consideration should be given to the control of static electricity in healthcare facilities including a review of 1990s recommendations in the light of more recent understanding and IEC 61340-6-1 requirements. Considerable benefits could be gained, including

- Reduction in infection rate due to electrostatic attraction of pathogens to equipment surfaces, patients, bed coverings and staff PPE.
- Reduction or elimination of nuisance electrostatic shocks to staff, patients and visitors to the facility.
- Reduction in upset, damage or data loss to electronic systems, including those providing critical functions to patients – this could reduce complications experienced by patients as well as provide cost savings from fewer repairs or less downtime of equipment.

• Reduction in risk of ignition of flammable atmospheres produced by products such as hand sanitizer gels.

In some cases, suitable materials, equipment or PPE may already be available on the market, supplying the needs of other industries or application areas. Where necessary, bespoke materials or suitable equipment for the healthcare sector might need to be developed.

Adaption of existing electrostatic spray technologies to the application of disinfectants in healthcare and related facilities could result in reduced disinfectant material consumption as well as improved coverage. The effective use of electrostatic sprays might, however, also depend to some extent on the adoption of IEC 61340-6-1 precautions to prevent electrostatic repulsion of charged disinfectant from the same polarity charged surfaces. Spray devices might also have to be modified and optimized for reliable electrostatic charging of spray particles with optimal diameter. Effective grounding of the spray unit and operating personnel should also be considered.

Deployment of electrostatic control measures would potentially increase the cost of equipment, materials, installations and consumables to some extent. It seems very likely, however, there would be practical financial savings accruing over many years in use, if fewer patient infections and reduced damage to electronic devices are achieved. Any benefit achievable in terms of reduced hospitalization times would be particularly beneficial to healthcare organizations, medical practitioners and patients alike. Electronic equipment is increasingly widespread in modern healthcare patient monitoring and medication dispensing and data records and management systems are held on computer equipment. Upset, data corruption and loss or computer and network crashes due to ESD can affect patient treatment or monitoring, or cause delays or other issues.

Any reduction achieved in the frequency of electrostatic shocks would no doubt be appreciated by patients, practitioners and other healthcare workers. These are often dismissed as a tolerable nuisance but can be surprisingly severe. Over more than 30 years in electrostatic consultancy, I have investigated several that have led to reported injury or formal disputes with the employer.

I look forward to a near future in which 'electrostatic hygiene' [19], including infection reduction by electrostatically optimal specification of materials used in the environment and in equipment, and compliance with IEC 61340-6-1, is as much an essential element of healthcare as handwashing and routine cleaning techniques have become today.

Acknowledgements

I would like to thank many colleagues for discussions over the years and during the writing of this chapter. In particular I would like to thank Dr James Matthews, Dr Dennis Henshaw and Dr Paul Holdstock for their assistance. I have endeavoured to use the best available information at the time of writing, but any inaccuracies or mistakes remain my own.

References

[1] Smallwood J.M. (2004) Static electricity in the modern human environment. *Electromagnetic Environments and Health in Buildings* Ed. D. Clements-Croome (eds). London: Spon Press.

[2] Health and Safety Executive. Chevron Pembroke amine regeneration unit explosion. Available from https://www.hse.gov.uk/comah/chevron-refinery. htm [accessed Aug. 2024]

[3] Wilson N. (1971) Some measurements of the static behaviour of carpets. *Report for the Federation of British Carpet Manufacturers* The Cotton, Silk and Man-Made Fibres Research Association.

[4] Cross J. (1987) *Electrostatics: Principles, Problems and Applications.* Bristol: IOP Publishing.

[5] International Electrotechnical Commission. Electrostatics – Part 1: Electrostatic phenomena – Principles and measurements. IEC TR 61340-1:2012 +AMD1:2020 CSV

[6] International Electrotechnical Commission. (2017) Explosive atmospheres Part 32-1. Electrostatic hazards, guidance. IEC TS 60079-32-1:2013 +AMD1:2017 CSV

[7] Energy Institute. (2001) Report on the risk of static ignition during vehicle refuelling. A study of the available relevant research. Available from https:// publishing.energyinst.org/topics/petroleum-product-storage-and-distribution/ filling-stations/report-on-the-risk-of-static-ignition-during-vehicle-refuelling.- a-study-of-the-available-relevant-research [accessed Jul. 2024]

[8] International Electrotechnical Commission. Electrostatics – Part 5-1: Protection of electronic devices from electrostatic phenomena – General requirements. IEC 61340-5-1:2024

[9] NHS Estates. (1996) *Health Guidance Notes. Static Discharges.*

[10] DHSS (1977) *Antistatic Precautions: Rubber, Plastics and fabric.* Hospital Technical Memorandum 1. HMSO. Out of print

[11] DHSS (1971) *Antistatic Precautions: Flooring in Anaesthetizing Areas.* Hospital Technical Memorandum 2. HMSO. Out of print

[12] Bray C.S. (1990) *Report of a Working Party to Review the Antistatic Requirements for Anaesthetizing Areas.* NHS Procurement Directorate. Department of Health.

[13] Holdstock P. and Wilson N. (1996), The effect of static charge generated on hospital bedding. *EOS/ESD Symposium Proceedings No. 18*, ESD Association, Rome, pp. 356–364.

[14] Kohani M. and Pecht M. (2017) Malfunctions of medical devices due to electrostatic occurrences. *IEEE Access* 6: 5805–5811.

[15] Viheriäkoski T. (2019) Electrostatic hazards in healthcare. *In Compliance.* Available from https://incompliancemag.com/electrostatic-hazards-in-health-care/ [accessed 15 Jul. 2024]

[16]   Velez K. Evaluating electrostatic sprayers for surface disinfection. Available from   https://www.cloroxpro.com/wp-content/uploads/2020/08/NI-51941_ CloroxPro_Electrostatic_Tech_White_Paper.pdf [accessed Jul. 2024]

[17]   National Audit Office (2000) The management and control of Hospital acquired infections in acute NHS Trusts in England. National Audit Office Press Notice.

[18]   International Electrotechnical Commission. Electrostatics – Part 6-1: Electrostatic control for healthcare – General requirements for facilities. IEC 61340-6-1:2018

[19]   Jamieson, I.A., Holdstock, P., ApSimon, H.M. and Bell, J.N.B. (2010) Building health: The need for electromagnetic hygiene? In *Institute of Physics Conference Series: Earth and Environmental Science.* 10 (1): 012007. Available from https://iopscience.iop.org/article/10.1088/1755-1315/10/1/ 012007/pdf [accessed 25 Feb. 2024]

[20]   Kubde D., Badge A.K., Ugemuge S., and Shahu S. (2023) Importance of hospital acquired infection control. *Cureus.* Available from https://www.cureus.com/articles/199727-importance-of-hospital-infection-control#!/ [accessed Aug. 2024]

[21]   ASHREA (2020) *ASHREA Handbook – HVAC Systems and Equipment.* Chapter 29

[22]   Mainelis G., Willeke K., Baron P., *et al.* (2001) Electrical charges on airborne micro-organisms. *Journal of Aerosol Science.* 32: 1087–1110.

[23]   McMurray P.H. (1985) Aerosol wall losses in electrically charged chambers. *Aerosol Science and Technology* 4: 249–268 DOI:10.1080/02786828508959054

[24]   Allen J.E., Wynne H., Ross F., Henshaw D.L., and Oakhill A. (2001) Biological aerosols, static charge and hospital infection. In: *Aerosols, Their generation, Behaviour and Applications, 12th Annual Conference of the Aerosol Society,* Bath University, 18–19 June 2001, pp. 99–102.

[25]   Allen J.E., Henshaw D.L., Wynne H., Ross F. and Oakhill A. (2003) Static electric charge may contribute to infections in bone marrow transplant wards. *Journal of Hospital Infection.* 54 (1): 80–81.

[26]   Allen J.E. (2005) Static electric fields as a mechanism of nosocomial infection. Presentation given at *Electric Fields and Discharges for Microbiology and Healthcare Applications,* Conference held by the Electrostatics Group of the Institute of Physics, London, 19 May 2005.

[27]   Allen J.E., Close J.J. and Henshaw D.L. (2006) Static electric fields as a mediator of hospital infection. *Indoor Built Environment.* 15 (1): 49–52.

[28]   Todd N.J., Millar M.R., Dealler S.R., and Williams S. (1990) Inadvertent intravenous infusion of Mucor during parenteral feeding of a neonate. *Journal of Hospital Infections.* 15: 295–297.

[29]   Becker R., Kristjanson A., and Walker J. (1996) Static electricity as a mechanism of bacterial transfer during endoscopic surgery. *Surgical Endoscopy.* 10: 397–399.

[30]   Cozanitis D.A., Ojajarvi J., and Makela P. (2008) Antistatic treatment for reducing airborne contamination of insulating materials in intensive care. *Acta Anaesthiologica Scandinavica.* 32 (4): 343–346.

[31]   Protective clothing – Performance requirements and tests methods for protective clothing against infective agents. EN 14126:2003.

[32]   Holdstock P. (2022) A review of electrostatic standards. *Journal of Electrostatics.* 115: 103562

[33]   Surgical clothing and drapes – Requirements and test methods – Surgical drapes and gowns. EN 13795-1:2019

[34]   Surgical clothing and drapes – Requirements and test methods – Clean air suits. EN 13795-2

[35]   Borland S. (2007) Hospital bans Crocs shoes over static risk. https://www.telegraph.co.uk/news/uknews/1562190/Hospital-bans-Crocs-shoes-over-static-risk.html. [accessed Jul. 2024]

[36]   International Standardisation Organisation. *Personal Protective Equipment – Safety Footwear.* ISO 20345

[37]   Council Directive 89/336/EEC of 3 May 1989 on the approximation of the laws of the Member States relating to electromagnetic compatibility. https://eur-lex.europa.eu/legal-content/en/ALL/?uri=CELEX:31989L0336 [accessed Aug. 2024]

[38]   Williams T. (2001) *EMC for Product Designers.* 3rd edn. Oxford: Newnes.

[39]   International Electrotechnical Commission. Electromagnetic compatibility (EMC) – Part 4-2: Testing and measurement techniques – Electrostatic discharge immunity test. IEC 61000-4-2:2008

[40]   Viheriäkoski T., Kokkonen M., Tamminen P., Kärjä E., Hillberg J., and Smallwood J. (2014) Electrostatic threats in hospital environment. *Proceedings of the EOS/ESD Symposium Paper 4B2.*

[41]   Medical electrical equipment. General requirements for basic safety and essential performance. Collateral Standard: Electromagnetic disturbances. Requirements and tests. EN60601-1-2

[42]   Robertson J.T. and Duong A. (2016) Electrostatic technology for surface disinfection in healthcare facilities. InfectionControl.tips Available from https://infectioncontrol.tips/2016/10/14/electrostatic-in-healthcare/ [accessed Aug. 2024]

[43]   Chauhan A., Patel M.K., Chaudhary S., *et al.* (2023) Efficacy evaluation of an air assisted electrostatic disinfection device for the effective disinfection and sanitization against the spread of pathogenic organisms. *Journal of Electrostatics* 123: 103107

[44]   Wood J.P., Magnuson M., Touati A., *et al.* (2021) Evaluation of electrostatic sprayers and foggers for the application of disinfectants in the era of SARS-CoV-2. *PLoS One* 16 (9): e0257434. https://doi.org/10.1371/journal.pone.0257434 [accessed 19 Jul. 2024]

[45]   NSC. Hand sanitizers can ignite, cause burns, experts warn workers. Available from   https://www.safetyandhealthmagazine.com/articles/19954-hand-sanitizers-can-ignite-cause-burns-experts-warn-workers [accessed Aug. 2024]

[46] NBC News (2013) Hand sanitizer may have ignited hospital fire that hurt girl, 11. https://www.nbcnews.com/health/health-news/hand-sanitizer-may-have-ignited-hospital-fire-hurt-girl-11-flna1C8433702 [accessed Aug. 2024]

[47] Patient Safety Authority (2006) Oxygen-enriched environments increase the fire risk from alcohol-based hand sanitizers. https://patientsafety.pa.gov/ADVISORIES/Pages/200612_11.aspx. [accessed Aug. 2024]

[48] International Electrotechnical Commission. Low-voltage Electrical installations – Part 7-710: Requirements for special installations or locations – Medical locations (2002–2021). IEC 60364-7-710

[49] International Electrotechnical Commission. Electrostatics – Part 2-1: Measurement methods – Ability of materials and products to dissipate static electric charge. IEC 61340-2-1:2015+AMD1:2022

[50] International Electrotechnical Commission. Electrostatics – Part 4-5: Standard test methods for specific applications – Methods for characterizing the electrostatic protection of footwear and flooring in combination with a person. IEC 61340-4-5:2018

[51] International Electrotechnical Commission. Electrostatics – Part 2–3: Methods of test for determining the resistance and resistivity of solid materials used to avoid electrostatic charge accumulation. IEC 61340-2-3:2016

[52] International Standardisation Organisation. Textiles – Test methods for evaluating the electrostatic propensity of fabrics – Part 2: Test method using rotary mechanical friction. ISO 18080-2

[53] International Standardisation Organisation. Textiles – Test methods for evaluating the electrostatic propensity of fabrics – Part 4: Test method using horizontal mechanical friction. ISO 18080-4

[54] International Standardisation Organisation. Textiles – Test methods for evaluating the – electrostatic propensity of fabrics – Part 3: Test method using manual friction. ISO 18080-3

[55] Torriani F., and Taplitz R. (2010) History of infection control. *Infectious Diseases* vol. 1, pp. 76–85. Amsterdam: Elsevier. doi:10.1016/B978-0-323-04579-7.00006-X [accessed Aug. 2024]

[56] British Standards Institution. Resilient, textile, laminate and modular multilayer floor coverings – Essential characteristics. BS EN 14041:2021

Chapter 7

The electromagnetic propulsion of future spacecraft

Andrew Michael Chugg[1]

7.1 Introduction

As a challenging objective, the definition of a propulsion system capable of taking a spacecraft crewed by humans to the nearby stars within a single human lifetime has been addressed. This necessitates a mean speed of travel at a substantial proportion of the speed of light, since even the very nearest star, Proxima Centauri, lies at a distance of 4.246 light years, and the journey should be reckoned as of the order of ten light years to bring a useful variety of neighbouring star systems within range: there are 32 other stars within 12.5 light years of the Sun [1]. The pursuit of this objective is particularly justified by the unattractiveness of crewed interstellar missions of such a long duration that generations of the crew are expected to spend their entire lives in transit. Consequently, this journey time of a few decades is necessary to make interstellar travel reasonably viable for humans.

7.2 Factors governing the speed achievable by spacecraft drives

There is a simple differential equation describing the variation of spacecraft mass M and speed v with time t assuming a propellant exhaust speed v_e and initial velocity $v_0 = 0$, initial overall mass $M = M_0$ and envisioning the velocity v accelerating to a maximum speed v_{max} at a constant rate of fuel mass ejection $dM/dt = k_M$. This expression (as with all expressions in this chapter) is non-relativistic for simplicity but provides a good approximation up to speeds around half the speed of light:

$$M\frac{dv}{dt} = (M_0 - k_M t)\frac{dv}{dt} = v_e\frac{dM}{dt} = v_e k_M \qquad (7.1)$$

[1]MBDA UK Limited, Bristol, UK

$$\frac{dv}{dt} = \frac{v_e k_M}{(M_0 - k_M t)} \tag{7.2}$$

On integrating with respect to time:

$$v = \int \frac{v_e k_M}{(M_0 - k_M t)} dt = \int \frac{v_e}{\left(\frac{M_0}{k_M} - t\right)} dt = v_e \ln \left|\frac{M_0}{k_M}\right| - v_e \ln \left|\frac{M_0}{k_M} - t\right| \tag{7.3}$$

This expression is only valid up to the maximum time t_{max} at which the fuel is exhausted. If φ is the fraction of the initial mass dedicated to acceleration fuel, then:

$$t_{max} < \varphi \frac{M_0}{k_M} \tag{7.4}$$

Hence, the maximum velocity v_{max} reached by the spacecraft may be defined as

$$v_{max} = v_e \ln \left|\frac{M_0}{k_M}\right| - v_e \ln \left|\frac{M_0}{k_M}(1 - \varphi)\right| = -v_e \ln |(1 - \varphi)| \tag{7.5}$$

Therefore, the speed achieved by the spacecraft depends only on the exhaust velocity of the fuel and the fraction of the spacecraft mass that is assigned to the acceleration fuel. This interdependency is graphed in Figure 7.1 with logarithmic axes and with the spacecraft mass fraction $(1 - \varphi)$ plotted instead of φ. The graph shows a relatively weak dependency on $(1 - \varphi)$, meaning that a very large fraction of the spacecraft mass needs to be dedicated to fuel before v_{max} can greatly exceed the exhaust velocity v_e. Furthermore, by symmetry, deceleration requires the same

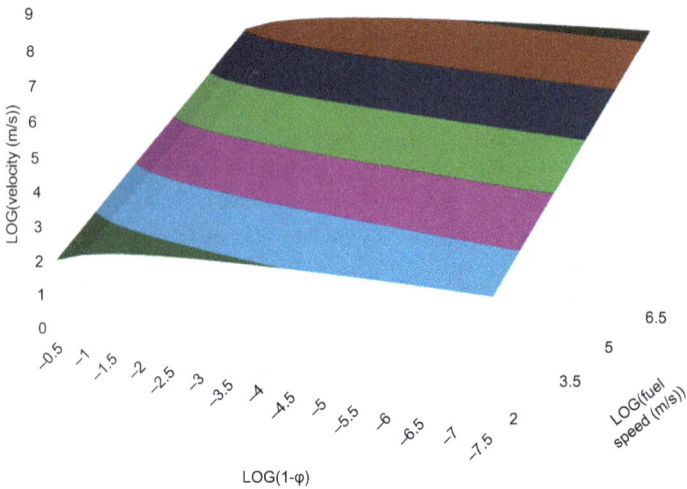

Figure 7.1   The dependency of spacecraft maximum velocity on fuel exhaust speed and fuel mass fraction

fuel fraction φ, so the remaining payload fraction of the mass is just $(1 - \varphi)(1 - \varphi)$. This makes it highly inefficient (costly in fuel mass) to achieve spacecraft velocities greatly in excess of the fuel exhaust velocity. For example, merely to achieve $v_{max} = 2v_e$ requires that the fuel constitute 98.2% of the mass of the spacecraft. Consequently, v_e is the crucial factor and in turn depends upon the kinetic energy that can be imparted to the exhausted fuel material (i.e. v_e^2).

7.3   Chemical engines with liquid or solid propellants

Of course, propelling spacecraft need not be based either wholly or partly on electromagnetic technology. Traditionally to date, it has instead relied on exothermic chemical reactions. However, the exhaust velocities achievable [2] are extremely modest in comparison with the speed of light at 300,000 km/s:

- 1.7–2.9 km/s for liquid monopropellants
- 2.9–4.5 km/s for liquid bipropellants
- 2.1–3.2 km/s for solid propellants

A light year is 9.467×10^{12} km, so visiting even the nearest star using chemical engines would take $\sim$100,000 years, which is quite impractical.

7.4   High exhaust velocity engines

Ion engines employ a jet of particles that are ionized and thereby accelerated across an electrical potential difference to achieve an arbitrarily high exhaust velocity. However, they tend to suffer from relatively low thrusts, and there remains the problem of how the energy to accelerate the ions is sourced.

In accordance with Einstein's mass-energy equivalence equation $E = mc^2$, the energy available to a conventional spacecraft is limited to the energy that can be released from the mass of its fuel. In effect, it is necessary to convert a large fraction of the fuel's mass to energy in order to achieve a maximum velocity that is a significant fraction of the speed of light. However, the most efficient practicable process for converting mass to energy is the nuclear fusion of hydrogen isotopes, which results in the conversion of only about 1% of the mass of the fusion fuel to energy. Neglecting any modest leverage from adopting a tiny value of $(1-\varphi)$, the effect is an absolute maximum spacecraft velocity of about $c/10$. This assumes a 100% efficiency of conversion of mass energy to spacecraft kinetic energy $Mc^2/100$, which is overly optimistic. Furthermore, due to the need to decelerate the spacecraft, only about half of the theoretical energy $Mc^2/200$ is available for the acceleration phase, yielding $v_{max} = c/14.14$. This suggests journey times to our neighbouring stars of around a century or more even with a fusion powered engine of near perfect efficiency.

It should be noted that a process that achieves 100% conversion of mass to energy, matter–antimatter annihilation, is known on the sub-atomic scale. However, neither scaling up antimatter production to the macroscopic level nor its long-term containment and storage are currently technologically practicable. Even if antimatter fuel were available, it would be devastatingly explosive. This recalls the early airships, which used hydrogen gas for lift and proved prone to disastrous detonations.

7.5   Light beams

In order to explore ways to circumvent the limitations on the performance of conventional fuel-driven spacecraft, it is necessary to appreciate the attractions of light beams as a means of propulsion. A more efficient or convenient propellant than electromagnetic radiation itself is hard to imagine, because photons intrinsically possess the highest possible velocity of any known propulsive material. Photons in space travel at virtually the speed of light in vacuum c. It follows that a collimated beam of photons emitted from a spacecraft or alternatively reflected from a thin-sheet mirror (a light-sail) will maximize the rate of change in its momentum relative to a jet of any other type similarly exhausted or reflected at the same rate of energy transfer by the jet. This derives directly from $E = mc^2$: a kilogram of photons travelling at c will always have more momentum than a kilogram of other matter travelling at some speed $v < c$, so long as the kilogram weight of such alternative matter also includes the part of its mass attributable to its kinetic energy.

Other attractive properties of light beams are the ease with which they may be generated and collimated by lasers and reflected and focused over large distances by curved mirrors or lenses.

This means that it is reasonable to consider the possibility of using a light sail to power a spacecraft instead of requiring that it carry all the energy for its propulsion within itself. Such an extraneous source of power brings propulsion to velocities at a larger fraction of the speed of light within the bounds of physical feasibility.

In fact, there is a laser-boosted light sail mission to Alpha Centauri that has been proposed called Breakthrough Starshot, but it is an ultra-lightweight, unmanned fly-by (only) mission with no deceleration [3]. An experimental light sail spacecraft called IKAROS was launched and successfully flown in 2010 using solar light to achieve a 1.12 mN thrust from a 196 m^2 light sail [4].

7.6   Light sources

Traditionally, it has been imagined that spacecraft utilizing light sails would exploit light sources already present in space: most particularly starlight and especially the light emissions of the Sun, whilst near enough to the Sun for it to be the dominant light source. At the orbit of the Earth, the power density in sunlight is 1353 W/m^2 in the radial direction [5]. Even in nearby interstellar space, where sunlight has ceased to be dominant, the energy density in starlight is about 0.45 eV/cm^3 [6]. The other significant electromagnetic radiation still present in interstellar space is the cosmic microwave background (CMB) with a color temperature of 2.7 K, but this can be calculated to have an energy density of only about a seventh of starlight.

Over very long periods of time, it could be possible to exploit naturally occurring light sources to accelerate a light sail up to a significant fraction of the speed of light. However, time is of the essence, if journeys to nearby stars are to be accomplished in decades, so the criterion should be to maximize the rate of acceleration of the light sail. Nevertheless, a human crew could not comfortably experience an acceleration much in excess of $g = 10$ m/s^2 over the long durations necessary to reach a significant fraction of c at an acceleration of g. For example, in non-relativistic kinematics it takes a time $t = 1.5 \times 10^7$ s = 0.475 years to reach $c/2 = 1.5 \times 10^8$ m/s at $g = 10$ m/s^2. In that time, the vessel will have travelled a distance $s = gt^2/2 = 1.125 \times 10^{12}$ km = 7520 astronomical units.

To achieve such a high acceleration over such a distance the light source needs to be a customized artificial beam with a high degree of collimation. This specification is only fulfilled by a large laser with a protracted coherence time and a mirror dish of immense diameter with a sensitively adaptive surface curvature to focus the light beam deep into interstellar space.

Over what range can a reflector dish of diameter d_A reflect the light incident upon it from the laser into the cross-section of a circular light sail of radius r_{sail} at a distance s? A fundamental limitation arises from the diffraction of the light beam out of the cross-section of the reflector dish (or also out of its own cross-section). The first minimum in the far field diffraction pattern for a disc occurs at a small angle θ defined in terms of d_A and the wavelength λ of the light:

$$\theta \approx \frac{1.33\lambda}{d_A} = \frac{r_{\text{sail}}}{s} \tag{7.6}$$

The need to avoid interference by the Earth's atmosphere and the very large scale of the structure may dictate that the laser be located in space in solar orbit or possibly attached to an airless body such as the Moon, Mercury, an asteroid or planetoid. If located in space, it will be necessary to tether the system to a massive body by some means, because of the large reactionary force upon the light generation apparatus itself. The possibilities include gravitational tethering in suborbital trajectories. The reactionary acceleration will be at least equal to the acceleration induced upon the light sail divided by the ratio of the mass of the beam-generating system to the mass of the spacecraft, yet probably significantly larger due to parts of the generated beam that miss the light sail. The light beam could be powered by solar radiation, perhaps by deploying a large array of solar cells in the vicinity of its laser. This may militate in favour of a physical location relatively closer to the Sun. The configuration is outlined in Figure 7.2.

7.7   Light sail design

In order to sail a pseudo-isotropic light source, such as starlight, it could be desirable to engineer a sail with significant anisotropy in its optical properties between

Figure 7.2   Laser beam forming and direction upon a spacecraft with a light sail

the front and rear surfaces. For example, making the front surface almost transparent whilst maintaining near perfect reflection over the relevant range of wavelength at the rear surface. Such properties can potentially be engineered using stacks of alternating high and low refractive index material or material engineered at the atomic level to present a gradient in its refractive index.

However, for the situation that a unidirectional, monochromatic beam is incident on the rear of the sail, a simple sheet of uniform metal with a highly polished rear beam-facing surface is likely to be optimal.

It needs to be a design aim that the light sail should dominate the mass of the spacecraft, so that the extra mass beyond the mass of the sail itself that the incident light flux needs to accelerate can be regarded as minimal. So long as the light sail dominates the spacecraft mass, the acceleration at a given normally incident light flux is inversely proportional to the mass thickness of the sail material, so it will be desirable to make the sail as thin as possible subject to retaining its mechanical integrity against any disruptive forces or instabilities.

If the light sail is envisaged to have a circular cross-section in the direction of incidence of the light beam, it would be unstable to crinkling as a thin, flat sheet. Just as with the wind sail of a seagoing vessel, the light sail needs to bow slightly such that the stress field generated by the central mass of the living accommodation and other non-sail-related parts on the sail is manifested as a circularly symmetrical degree of strain. The result should be a shallow dome configuration for the light sail, convex facing backwards towards the incoming light beam (but the light sail would be concave backwards, if the ancillary mass were distributed around its periphery). Furthermore, a supportive structure of hollow cylindrical struts is likely to be needed on the forward-directed concave side of the sail to maintain structural integrity against instabilities and to ensure that the forces from the incident light beam pressure are distributed evenly through the structure.

The struts need to be single-crystal hollow tubes for maximum strength and rigidity per mass. Normal metals are polycrystalline structures with significant weakness at the crystal boundaries and consequential reduction of strength and stiffness. The same single-crystal structure for the sail material itself will aid in minimizing its thickness and the degree of support it requires from the struts. In order to produce such a perfect micro-structure in the struts and the sail, an epitaxial three-dimensional printing process is likely to be required. It will be a technological challenge to produce such perfectly crystalline material on the potentially enormous scale of such a spacecraft.

7.8   Light sail performance

For a perfectly reflective surface, the radiation pressure P_{rad} exerted by electromagnetic radiation of intensity I in W/m^2 is given by:

$$P_{rad} = \frac{2I}{c} \tag{7.7}$$

where c is the speed of light.

For a notionally circular and shallowly domed light reflector sail of thickness δ and density ρ, the light sail mass per area is $\delta\rho$.

Hence for an acceleration of 1g:

$$P_{rad} = \frac{2I}{c} = g\delta\rho \tag{7.8}$$

Therefore, minimizing the required light intensity requires that both the thickness and density of the sail material be minimized.

However, a limitation on how thin the sail can be and on the nature of its substance arises from the degree of penetration of the light beam into the material. In effect, if the sail is too thin, it becomes transparent to the incident light beam to some extent. This transparency is governed by the skin depth d_{skin} of the light penetration at the selected frequency f in a material of conductivity σ and permittivity ε and permeability μ:

$$d_{skin} = \sqrt{\frac{1}{\pi f \mu \sigma}} \sqrt{\sqrt{1 + (2\pi f \epsilon/\sigma)^2} + 2\pi f \epsilon/\sigma} \tag{7.9}$$

For a good conductor such as a metal:

$$d_{skin} \approx \sqrt{\frac{1}{\pi f \mu \sigma}} \tag{7.10}$$

Evidently, f should be as large as possible, yet not so large that the photons are individually capable of ionizing the substance, which could destroy the sail. In particular, an especially high conductivity is implied and that probably restricts the choice to a metal as indeed is still the most usual choice for the silvering of mirrors. The skin depth is the distance for a $1/e$ decay of the electric field strength with depth into the material, so a light sail would need to be at least a few skin depths thick in order to exhibit the high reflectivity required of a mirror.

It might be queried whether a superconducting material is a possible choice with the objective of realizing a vanishingly small skin depth? The temperature of interstellar space could be as low as the temperature of the CMB, and many substances have a superconducting transition temperature above 2.7 K. However, superconductivity is not currently realizable either in very thin layers or at frequencies as high as the frequencies of visible light photons: high-frequency oscillating surface currents occur below the London depth and the Cooper pairs that produce superconductivity are disrupted.

7.9   Light sail cooling

The penetration of the light beam up to a few skin depths into the sail means that there will be some Ohmic heating in the surface layer of the sail. It will be necessary for the consequential temperature rise in the sail to reach an equilibrium significantly below the melting point of the sail material, or else the sail would be destroyed by the light beam. The only means of cooling that would readily be available is pseudo-blackbody radiation to space according to the Stefan–Boltzmann law.

The Maxwell equations and the Poynting vector derived from them dictate a relationship between the incident power density I in the light beam and surface current induced in the light sail. First, power density is proportional to the square of the rms E-field E or H-field H where the permittivity of free space is $\varepsilon_0 = 8.854 \times 10^{-12}$ F/m and the permeability of free space is $\mu_0 = 1.25663706 \times 10^{-6}$ H/m:

$$I = \sqrt{\frac{\epsilon_0}{\mu_0}}E^2 = \sqrt{\frac{\mu_0}{\epsilon_0}}H^2 \tag{7.11}$$

The current density per distance in the direction parallel to the H-field across the sail surface flowing in the direction of the electric field vector is J_d and:

$$J_d = H \tag{7.12}$$

However, this surface current J_d actually penetrates a few skin depths into the light sail, so the complete behaviour can be described by a current density by cross-sectional area J, which is a decaying exponential function of the depth x into the sail surface such that:

$$J_d = \int_0^\infty J dx = \int_0^\infty J_0 e^{-x/d_{skin}} dx = \left[J_0 d_{skin} e^{-x/d_{skin}} \right]_\infty^0 = J_0 d_{skin} \tag{7.13}$$

The rate of heat energy Q dissipated per unit surface area of sail due to the resistivity $\rho_e = 1/\sigma$ (reciprocal of conductivity) of the sail material is therefore:

$$Q = \int_0^\infty J^2 \rho_e dx = \int_0^\infty \frac{J_0^2 e^{-2x/d_{skin}}}{\sigma} dx = \left[\frac{J_0^2 d_{skin} e^{-2x/d_{skin}}}{2\sigma} \right]_\infty^0 = \frac{J_0^2 d_{skin}}{2\sigma} \quad (7.14)$$

In equilibrium, the sail must rise to a temperature Θ, so as to radiate all the deposited heat, giving:

$$Q = \frac{J_0^2 d_{skin}}{2\sigma} = 2s_B \Theta^4 = \frac{H^2}{2\sigma d_{skin}} = \frac{I}{2\sigma d_{skin}} \sqrt{\frac{\epsilon_0}{\mu_0}} \quad (7.15)$$

where s_B is the Stefan–Boltzmann constant 5.67×10^{-8} W m^{-2} K^{-4}, and the radiated heat is doubled on account of contributions from both the front and rear faces of the sail. Evidently, this temperature is independent of the thickness δ of the sail so long as it is at least several skin depths thick:

$$\delta > 3d_{skin} \quad (7.16)$$

Hence, the peak power density irradiating the sail is proportional to the electrical conductivity of the sail material, its skin depth and the fourth power of the maximum allowable absolute sail temperature. Note however that the conductivity and therefore the skin depth are also functions of the temperature. It has also been assumed that the temperature variation $\Delta\Theta$ across the thickness of the sail is negligible compared to its absolute temperature Θ. This criterion additionally depends upon the thermal conductivity k such that

$$\frac{\Delta\Theta}{\Theta} \approx \frac{Q\delta}{k\Theta} \ll 1 \quad (7.17)$$

So a material of high thermal conductivity is desirable, again indicating a metal. A high melting point will also be very helpful given the high degree of leverage from maximizing the operating temperature.

7.10   The problem of deceleration in interstellar space

A difficulty with using a light beam based in our solar system to accelerate a light sail spacecraft into interstellar space is that there is no suitable light beam available to decelerate the spacecraft as it approaches another star system. It would eventually be possible to build a duplicate laser light beam generator in the destination system by sending a fusion-powered conventional spacecraft mission, but that would potentially take centuries. Is there any other alternative means of decelerating a light sail spacecraft from travelling at a substantial fraction of the speed of light?

The difficulty in decelerating in a vacuum is that it is empty, so there is a lack of a medium to which to transfer the kinetic energy of the vessel. However, interstellar space is not altogether empty, although the density of atoms is rather low and significantly variable from one place to another.

In general, interstellar matter can be described as highly inhomogeneous and far from thermal equilibrium. The diffuse medium, dominated by hydrogen and protons, exhibits particle densities of 0.01–1 cm^{-3} in the ionized phase and about 1–10^3 cm^{-3} in the cooler neutral phase [7,8].

The question may therefore be posed of whether the drag from collisions of the light sail with these very low number densities of protons and hydrogen molecules can significantly decelerate the sail?

Starting from Newton's second law, the force $F = ma$, where a is the deceleration and $m = A\delta\rho_{sail}$ such that A is the surface area of the sail, δ is its thickness and ρ_{sail} is the density of the sail material. The kinetic energy dE acquired by the hydrogen atoms at density ρ_{space} swept up in the distance dx as they are assumed to be captured into the sail is the force F exerted on the sail through the distance x:

$$dE = Fdx = \rho_{space}Av^2dx/2 \tag{7.18}$$

$$a = \frac{F}{m} = \frac{1}{A\delta\rho_{sail}}\frac{dE}{dx} = \frac{1}{A\delta\rho_{sail}}\rho_{space}A\frac{v^2}{2} = \frac{\rho_{space}v^2}{2\delta\rho_{sail}} \tag{7.19}$$

If the acceleration from laser illumination of the sail is to be of the order of 1 g (10 m/s^2), the deceleration needs to be at least of the order of g/10, when v is a substantial proportion of the speed of light c.

The mass of a hydrogen nucleus is $1.6735575 \times 10^{-27}$ kg, so, for example, at 1 atom per cubic centimetre $\rho_{space} = 1.6735575 \times 10^{-21}$ kg/m^3. Let it be supposed for illustrative purposes that the sail material is tantalum, of which the density is 16,600 kg/m^3. (It shall subsequently transpire that metals with superconducting transition temperatures above the temperature of the CMB at 2.7 K are of particular interest and tantalum has a superconducting transition temperature of 4.48 K.) Considering the deceleration, supposing all the incident protons to be arrested, when $v = c/2 = 1.5 \times 10^8$ m/s, the sail thickness δ may be calculated from (7.19) for the case that the interstellar matter deceleration is g/10 = 1 m/s^2:

$$\delta = \frac{\rho_{space}v^2}{2a\rho_{sail}} = \frac{10\rho_{space}c^2}{8g\rho_{sail}} \approx 1.132\text{nm} \tag{7.20}$$

The interatomic (nearest neighbour) distance in metallic tantalum is about 0.285 nm, so this suggests a sail no more than four atoms thick or less before the deceleration from interstellar hydrogen capture becomes significant [9].

Integrating to find the velocity history during such a deceleration:

$$-\int\frac{dv}{v^2} = \int\frac{\rho_{space}}{2\delta\rho_{sail}}dt = k + \frac{1}{v} = \frac{\rho_{space}}{2\delta\rho_{sail}}t \tag{7.21}$$

$$v = \frac{1}{\frac{1}{v_0} + \frac{\rho_{space}}{2\delta\rho_{sail}}t} \tag{7.22}$$

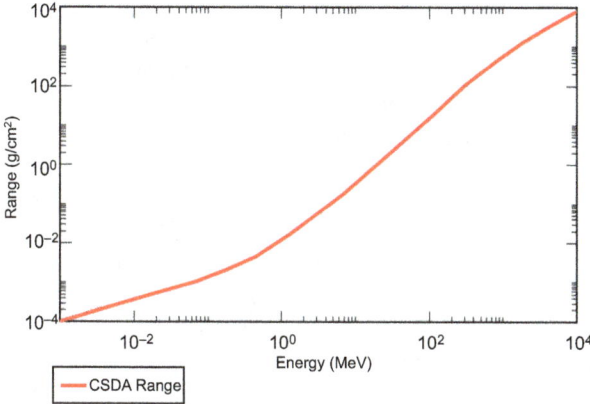

Figure 7.3   Range in mass thickness of protons of a range of energies in tantalum (continuous-slowing-down-approximation)

Integrating again to determine the distance travelled during the deceleration (such that $x = 0$ when $t = 0$):

$$x = \int \frac{dt}{\frac{1}{v_0} + \frac{\rho_{space}}{2\delta\rho_{sail}}t} = \left[\frac{2\delta\rho_{sail}}{\rho_{space}} \ln\left|\frac{1}{v_0} + \frac{\rho_{space}}{2\delta\rho_{sail}}t\right| \right]_0^\tau \tag{7.23}$$

$$x = \frac{2\delta\rho_{sail}}{\rho_{space}} \left(\ln\left|\frac{1}{v_0} + \frac{\rho_{space}}{2\delta\rho_{sail}}\tau\right| - \ln\left|\frac{1}{v_0}\right| \right) \tag{7.24}$$

However, x does not converge to a finite distance as the duration τ increases, but instead perpetually increases. This is because this form of deceleration is proportional to the square of the speed, so it weakens very considerably at low speeds. Consequently, some additional type of deceleration would need to be introduced at low speeds in order to bring the vessel to a complete halt.

A more serious flaw within this concept is that the range of protons in tantalum at the relevant velocities is far too great for their motion relative to the sail to be arrested within the thickness of the sail. The range of protons in tantalum is shown graphically in Figure 7.3 [10]. The kinetic energy of a proton at $c/2$ is 118 MeV, so its range in tantalum is of the order of 20 g/cm^2 or 1.2 cm. Clearly, there is a huge disparity of the order of eight orders of magnitude between the thickness of a sail that could be decelerated sufficiently by arresting impinging interstellar protons and the thickness required actually to arrest them. That is to say that the interstellar medium is simply far too thin for this species of drag to form a useful deceleration mechanism.

7.11   Exploiting magnetospheric drag

In order to generate a useful amount of deceleration, the light sail spacecraft would need to present a much larger cross-section to the interstellar medium than the area of its sail, and yet, it would need to arrest or significantly deflect the interstellar protons with a high degree of efficiency. There is a possible means of achieving these ostensibly conflicted aims with reference to the behaviour of planetary magnetospheres in the solar wind.

A magnetic field is generated by currents circulating deep with the Earth and some other planets. Above the surface of the Earth, the form of this 'geomagnetic' field is not greatly different from the field of a magnetic dipole (small current loop) of magnetic dipole moment 8.22×10^{22} A m^2 located near the centre of the Earth and usually oriented at a modest angle to the rotational axis of the Earth. This field would extend infinitely into the reaches of space were it not for a stream of ionized particles emanating from the Sun known as the solar wind.

Ionized particles experience a centripetal force in proportion to the component of their velocity at right angles to magnetic field lines. They therefore execute a helical trajectory with its axis oriented parallel to the field lines. In rotating around a bunch of field lines, they themselves constitute a current loop, which generates a magnetic field diametrically opposed to the field lines that are causing them to loop the loop. Where the number density of the ionized particles in the solar wind exceeds a sufficient magnitude, they generate an opposing field that completely cancels the geomagnetic field. Conversely, closer to the Earth, where its field is stronger such that it cannot be cancelled by the solar wind, the curvature of the trajectories of solar wind particles tends to deflect them around the Earth, thus largely excluding the solar wind from regions of space sufficiently close to the Earth. The net effect is to create a surface called the magnetopause (Figure 7.4) facing into the solar wind, beyond which the geomagnetic field ceases to exist. Within the magnetopause, the geomagnetic dipole field lines are distorted so as to remain within the magnetopause. Beyond the magnetopause, the interplanetary magnetic field is disrupted, and the solar wind particles are deflected as soon as they reach another surface called the bow shock, which is the sudden start of the disruption.

Of particular interest for the present account, the deflection of the solar wind between the bow shock and the magnetopause causes a ram pressure to be exerted upon the geomagnetosphere in the direction of the flow of the solar wind. If it were possible to set up a sufficiently strong magnetic dipole field around a spacecraft passing through interstellar space at some high velocity v, a similar magnetosphere would be formed around the vessel, and the ions (predominantly protons) present would exert a like pressure leading to a decelerating drag upon the spacecraft. In principle, this artificial magnetosphere could be very large by virtue of the low number density of ions in interstellar space, thus compensating for the thinness of the material by interacting with the interstellar medium over a vast cross-section.

For any magnetospheric structure, the ram pressure P_{ram} may be approximated [11] in terms of the velocity v of the ionized particles and the mass m_p of the particles and their number density n_p:

$$P_{ram} = m_p n_p v^2 = \rho_{space} v^2 \qquad (7.25)$$

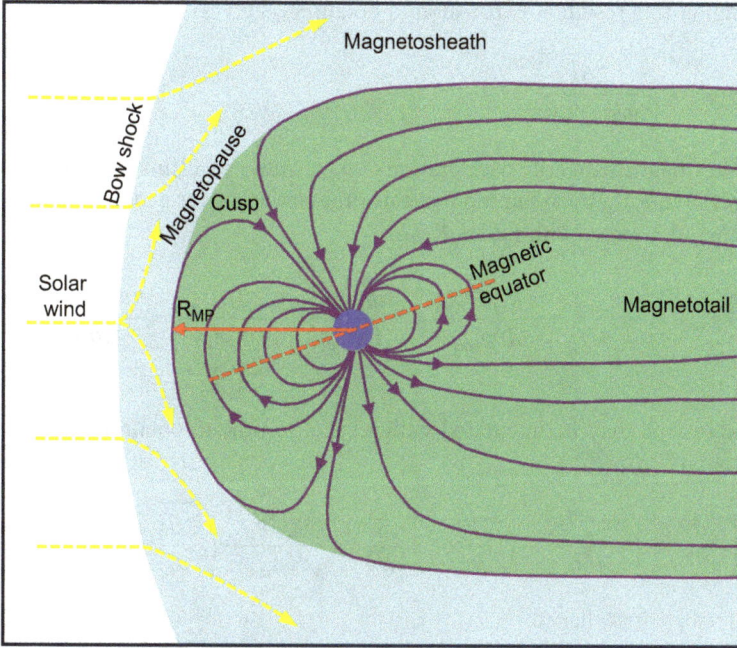

Figure 7.4   The Earth's magnetosphere

In order to infer the actual force exerted, it is necessary to estimate the radius R_{MP} of the magnetopause from the centre of the magnetic dipole in the direction of the spacecraft's velocity v through the interstellar medium.

The condition defining this radius [12] is that it is where the dynamic ram pressure from the ionized particle number density is equal to the magnetic pressure from the magnetospheric field:

$$P_{ram} = \rho_{space}v^2 \approx \frac{2B(r)^2}{\mu_0} \tag{7.26}$$

where $B(r)$ is the magnetic field strength of the magnetosphere in SI units (B in Tesla, μ_0 in H/m) at radius r. Since a dipole magnetic field strength varies with radius as $1/r^3$, the magnetic field strength can be written as:

$$B = \frac{B_0}{r^3} \tag{7.27}$$

where B_0 is the magnetosphere's magnetic magnitude parameter in units of Tm^3. Hence,

$$\rho_{space}v^2 \approx \frac{2B_0^2}{\mu_0 R_{MP}^6} \tag{7.28}$$

Rearranging this resultant expression to obtain R_{MP} gives:

$$R_{MP} \approx \sqrt[6]{\frac{2B_0^2}{\mu_0 \rho_{space} v^2}} \tag{7.29}$$

Now, the magnetospheric deceleration $-a_{mag}$ may be estimated from the ram pressure force F_{mag} by assuming that this pressure acts over a cross-section, which is the area of a circle with radius R_{MP}:

$$-a_{mag} = \frac{-F_{mag}}{m} = \frac{\pi R_{MP}^2 P_{ram}}{A\delta\rho_{sail}} = \frac{\pi m_p n_p v^2}{A\delta\rho_{sail}} \sqrt[3]{\frac{2B_0^2}{\mu_0 m_p n_p v^2}} = \frac{\pi m_p^{2/3} n_p^{\frac{2}{3}} v^{4/3}}{A\delta\rho_{sail}} \sqrt[3]{\frac{2B_0^2}{\mu_0}} \tag{7.30}$$

This expression may be integrated with respect to time to obtain the velocity history using $a = dv/dt$:

$$\int \frac{-dv}{v^{4/3}} = \int \frac{\pi m_p^{2/3} n_p^{2/3}}{A\delta\rho_{sail}} \sqrt[3]{\frac{2B_0^2}{\mu_0}} dt = \frac{3}{\sqrt[3]{v}} = \frac{\pi m_p^{2/3} n_p^{2/3}}{A\delta\rho_{sail}} \sqrt[3]{\frac{2B_0^2}{\mu_0}} t + k \tag{7.31}$$

When $t = 0$, $v = v_0$, hence:

$$\frac{3}{\sqrt[3]{v_0}} = k \tag{7.32}$$

Therefore,

$$v = \frac{1}{\left(\frac{\pi m_p^{2/3} n_p^{2/3}}{3A\delta\rho_{sail}} \sqrt[3]{\frac{2B_0^2}{\mu_0}} t + \frac{1}{\sqrt[3]{v_0}} \right)^3} \tag{7.33}$$

The mass of a proton m_p is a fixed parameter and the number density of free protons in interstellar space n_p is (initially at least) a given characteristic of the particular region of space. Furthermore, the spacecraft mass, assumed to be dominated by the light sail mass $A\delta\rho_{sail}$, is already considered to be minimal for the purpose of defining an optimal light sail for the acceleration phase. Consequently, effectuating a sufficiently fast deceleration is largely a question of creating a magnetic field with a sufficiently large magnitude parameter B_0.

This parameter is proportional to the magnitude M_{mag} of the magnetic moment vector $\mathbf{M_{mag}}$ of the magnetosphere, which is the product of the loop current magnitude i with the area of the current loop A_M with a direction at right angles to the plane of the loop in the sense as shown in Figure 7.5.

$$B_0 = \mu_0 M_{mag} \tag{7.34}$$

Essentially, the task is to create a current loop with as large a current as possible and as large an area as possible. But first and foremost, the current must be

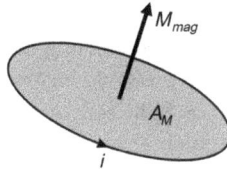

Figure 7.5   Definition of the magnetic moment vector

maintained for a long period (months-years) with minimal dissipation of energy, for example, in the form of resistive heating of the conducting ring carrying the current. This consideration militates in favour of fabricating the annular current conductor from superconducting material.

Logically, the annular current ring needs to be fabricated from the material of the light sail since that constitutes the great majority of the material available. It has already been noted that a superconducting material with a transition temperature above the 2.7 K temperature of the CMB is desirable, lest the cosmic microwave background should heat the ring above its superconducting transition temperature. Furthermore, it has been concluded that metal is the best type of material for the light sail since it needs to be an excellent electrical conductor and a good thermal conductor with strong interatomic bonds. Candidate metals with a transition temperature greater than 2.7 K would include tin (T_c = 3.72 K) and tantalum (T_c = 4.48 K) [13].

The deflection of interstellar ions is probably most efficient if the magnetic moment vector is oriented at right angles to the direction of motion, so a 90° rotation relative to the light sail orientation is implied.

The maximum current is constrained by an empirical limit on the current per cross-sectional area in a superconductor. This appears to be the practical limit on the dipole strength because superconductors lose their unique superconducting properties above a critical current density J_c of about 10^5 A/cm^2 according to [14,15]

> There is a limit to the maximum magnetic field, and thus there is a maximum current that can flow through a superconductor. This limit, called the critical field, is the maximum magnetic field where the superconductor still remains superconducting. The corresponding critical current, I_c, is the maximum current that can be passed through the superconductor before it reverts to normal conduction.

See also some examples of critical current densities around 20 kA/cm^2 for real materials at [16].

For a fixed mass of metal available to form the superconducting ring, the ring's cross-section and, therefore, the current limit will scale in inverse proportion to the radius. However, the area of the current ring increases as the square of the ring radius. Therefore, the overall magnetic moment scales in proportion to the radius, and it is clear that the maximum feasible radius is optimal.

The forging of the ring therefore implies the dismantling of the sail sheet elements and processing them into superconducting current ring sections to be assembled into an enormous circular loop probably beyond the radius of the sail rim.

Defining the volume V of the light sail of radius r_{sail} and thickness δ to be equal to the volume of the superconducting annulus of radius r_{ring} and of cross-section S:

$$V = \pi\delta r_{sail}^2 = 2\pi r_{ring}S \rightarrow S = \frac{\delta r_{sail}^2}{2r_{ring}} \tag{7.35}$$

This allows B_0 to be defined in terms of the critical current density J_c and the sail radius and thickness:

$$B_0 = \mu_0 M_{mag} = \pi\mu_0 J_c S r_{ring}^2 = \frac{\pi\mu_0 J_c \delta r_{ring} r_{sail}^2}{2} \tag{7.36}$$

This can be used to substitute for B_0 in (7.33):

$$v = \frac{1}{\left(\sqrt[3]{\frac{\pi^2\mu_0 r_{ring}^2 J_c^2 m_p^2 n_p^2}{54 r_{sail}^2 \delta \rho_{sail}^3}}t + \frac{1}{\sqrt[3]{v_0}}\right)^3} \tag{7.37}$$

It is notable that the deceleration depends on the ratio of the ring radius to the sail radius and increases as this ratio increases. The deceleration also increases as the density of ions (protons) increases, but the deceleration decreases if the thickness of the sail or its density increases.

Some realistic values for the parameters in (7.37) may be proposed in order to investigate the efficacy of this method of braking:

$J_c < 1\times10^9$ A/m^2
$m_p = 1.67262192\times10^{-27}$ kg
$n_p < 1\times10^6$ m^{-3}
$\rho_{sail} = 16,600$ kg m^{-3} (tantalum)
$\mu_0 = 1.25663706\times10^{-6}$ m kg s^{-2} A^{-2}
$\delta = 2.2\times10^{-8}$ m (3 skin depths for tantalum at 6×10^{14} Hz)
$r_{ring}/r_{sail} = 5$
$v_0 = c/2 = 1.5\times10^8$ m/s

$$v = \frac{1}{\left(5.426x10^{-12}\, t + 0.001882\right)^3} \quad (t \text{ in seconds}) \tag{7.38}$$

$$v = \frac{1}{\left(0.000171\, t + 0.001882\right)^3} \quad (t \text{ in years}) \tag{7.39}$$

The result is shown in Figure 7.6. The magnetospheric drag mechanism seems promising in that it could reasonably achieve a deceleration from half the speed of light to a tenth of the speed of light in about 7.8 years.

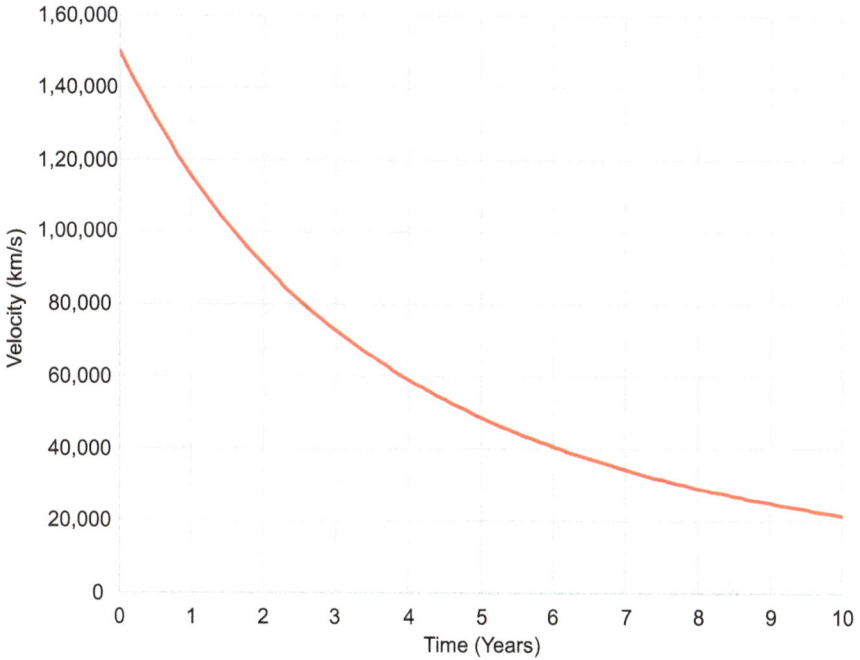

Figure 7.6   Example case of deceleration exploiting magnetospheric drag

Magnetospheric drag not only provides a vastly stronger deceleration than the drag on the light sail, but it also holds up better at lower spacecraft velocities relative to the interstellar matter. This is because the magnetic bubble around the spacecraft expands as its velocity decreases and this partially compensates for the decreasing incidence velocity of the interstellar matter. Nevertheless, it still does not achieve zero velocity, because the deceleration becomes slow at low spacecraft velocity such that the velocity never reaches zero in a finite time.

As the spacecraft slows, it is therefore necessary for an ancillary deceleration mechanism to be added in or to take over at low velocity to completely arrest the spacecraft: for example, a fusion drive. At $c/10$, the kinetic energy of the spacecraft is only about 0.5% of the mass energy bound up in its substance, so an efficient fusion drive becomes practicable as a means of completing the deceleration. The life-support, fusion fuel and fusion engine functions of the spacecraft would detach from the magnetospheric deceleration ring at the start of this phase so as to continue the deceleration in isolation from the ring in order not to need to find the energy to decelerate the ring as well.

The cost of generating the magnetospheric deceleration is the energy that must be transferred into the magnetic dipole field. To quantify this, it is necessary to define the energy stored in the magnetic field of a loop current as a function of its magnetic moment. It is a standard result that the energy E_L stored in the magnetic field of an inductance L in which a current I is circulating is given by:

$$E_L = \frac{LI^2}{2} \tag{7.40}$$

It is also a standard result that the inductance of a circular loop of current-carrying wire [17] is given by:

$$L = N^2 \mu_0 \mu_r R \left[\ln\left(\frac{8R}{w}\right) - 2 \right] \tag{7.41}$$

L = inductance (H)
R = loop radius (meters) = r_{ring}
w = wire radius (meters)
N = number of turns (1 for a single loop)
μ_0 = permeability of free space ($1.25663706 \times 10^{-6}$ m kg s^{-2} A^{-2})
μ_r = relative permeability (1 for free space)

$$w = \sqrt{S/\pi} \tag{7.42}$$

Hence,

$$L = \mu_0 r_{ring} \left[\ln\left(\frac{8r_{ring}}{\sqrt{S/\pi}}\right) - 2 \right] \tag{7.43}$$

Therefore,

$$E_L = \frac{\mu_0 r_{ring} J_c^2 S^2}{2} \left[\ln\left(\frac{8r_{ring}}{\sqrt{S/\pi}}\right) - 2 \right] \tag{7.44}$$

Substituting for S using (7.35):

$$E_L = \frac{\mu_0 J_c^2 \delta^2 r_{sail}^4}{8r_{ring}} \left[\ln\left(\frac{8r_{ring}\sqrt{2\pi r_{ring}/\delta}}{r_{sail}}\right) - 2 \right] \tag{7.45}$$

There needs to exist an engine of some kind to inject the current into the super-conducting ring, whether by direct injection or by induction. Once the current density has achieved its limiting value J_c, this engine merely needs to apply a holding force as necessary to prevent the current in the ring from diminishing. In principle, such a force need not inject any further energy, although in practice there are likely to be some losses, which must nevertheless be minimized.

For a planetary magnetosphere, there exists a ring of field lines that intersect the planet's surface around the magnetic poles in both the northern and southern hemispheres such that these same field lines pass through the surface of the magnetopause. That is to say, there are holes in the magnetopause above the magnetic poles. This means that solar wind particles from the surface of the magnetopause can travel down these holes to intersect the planetary atmosphere in rings around both magnetic poles. On Earth, this is the phenomenon that gives rise to the aurorae at high latitudes in both hemispheres due to the interaction of these solar wind particles with the high atmosphere. The analogous process for the magnetospheric spacecraft leads to the funnelling of a proportion of the interstellar protons into part of the area within the superconducting current ring, prospectively into small circular regions thereabouts. Consideration might be given to capturing these particles as a potential means of augmenting the fusion fuel reserves of the spacecraft as a useful by-product of the magnetospheric deceleration drive.

> One ring to tow them all, one ring to mind them,
> One ring to slow them all and in the darkness bind them.

7.12   The ionization of paths through neutral hydrogen

It has been noted that there are neutral species (mainly hydrogen) in interstellar space with typically a hundred times the density of the ionized species. It can also be seen with reference to (7.37) that the rate of deceleration of an artificial magnetosphere is sensitively dependent on the ambient ion number density with a positive correlation. This prompts the question of whether the ion density in the path of a spacecraft can be increased by ionizing some or all of the neutral phase material? The only likely means of achieving this would be a laser beam directed from our solar system. The laser light would need to be tuned to the ionization energy of hydrogen, which is 13.6 eV or 2.176×10^{-18} J, a photon frequency of 3.28×10^{15} Hz and a wavelength of 9.135×10^{-8} m, which is well into the ultraviolet. Such short wavelength lasers, such as excimer lasers, do exist, but it would be challenging to scale them up to the requisite size and power. Most particularly, the increased ionization is required near the destination star. Keeping the laser beam sufficiently focused (narrow) and targeted (aligned with the light sail path) over such a huge distance would be extremely difficult.

7.13   Parameters of a realistic spacecraft

An illustration of the quantitative values of the various parameters of the spacecraft and its propulsion systems can be based on a definition of the mass and material for the light sail and its payload. It is necessary for the payload to fulfill every other function of the spacecraft apart from the light sail itself including living accommodation and life support for a human crew for some decades. It would be hard to fulfill this purpose with less than around 100,000 tonnes (10^8 kg) of mass. However, this payload has been assumed to be of negligible mass relative to the mass of the light sail itself, which means that the total mass must be at least around a factor of ten larger at circa 1 million tonnes (10^9 kg).

   The metal of the sail needs to have a superconducting transition temperature above 2.7 K in order to keep the current ring superconducting. Tin is a possibility, but it has a melting temperature of only 231.9 °C. The heating effects of the laser light in the skin depth would risk melting a tin sail. Therefore, tantalum with an even higher superconducting transition temperature of 4.47 K appears to be a better candidate. Relevant parameters of tantalum include a density of 16,600 kg m^{-3} and electrical conductivity of 7.7 × 10^6 S/m at room temperature; in particular, it has an exceptionally high melting point of 3015 °C or 3288 K [18].

Choosing a laser wavelength of 500 nm and a frequency of 6 × 10^{14} Hz, the skin depth of tantalum may be calculated from (7.10) giving d_{skin} = 7.4 × 10^{-9} m. In accordance with (7.16), $\delta = 3d_{skin}$ = 2.2 × 10^{-8} m.

The next step is to define the light sail radius for the light sail using:

$$m = \pi \rho_{sail} \delta r_{sail}^2 \tag{7.46}$$

Hence, r_{sail} = 934 km.

The power and power density of the laser beam may be calculated for an acceleration of 1g using (7.7) and (7.8)

$$I = \frac{g\delta\rho c}{2} = 547 \text{kW}/\text{m}^2 \tag{7.47}$$

Therefore, the minimum total power in the beam P_{min} may be calculated on the assumption that virtually all the power is focused on the cross-section of the light sail:

$$P_{min} = \pi I r_{sail}^2 = 1.5 \times 10^{15} \text{W} \tag{7.48}$$

This is equivalent to the power released by the detonation of a nuclear bomb of 1 megaton yield every three seconds. This is engineering on a stupendous scale.

It is necessary to check that the heating from the penetration of the incident light into the skin depth of the light sail mirror surface leads to an equilibrium temperature in the sail that is well below its melting point. By adapting (7.15) with ε_0 = 8.854 × 10^{-12} F/m and the permeability of free space of μ_0 = 1.25663706 × 10^{-6} H/m with the Stefan–Boltzmann constant 5.67×10^{-8} W m^{-2} K^{-4}:

$$\Theta^4 = \frac{I}{4s_B \sigma d_{skin}} \sqrt{\frac{\varepsilon_0}{\mu_0}} \tag{7.49}$$

The result is that the equilibrium temperature is Θ = 579 K. This is confirmed to be well below the melting point of tantalum at 3288 K.

Given the assumption that the radius of the magnetospheric current ring is five times the radius of the light sail, it is now possible to calculate the wire cross-section and radius and thereby also the current and the energy stored in the artificial magnetosphere. First, using (7.35), the cross-section S of the current loop wire is determined to be 2.1 × 10^{-3} m^2, yielding a wire radius of 2.6 cm. Multiplying this cross-section by the limiting current density J_c defines a ring current of 2,100,000 A in the loop.

Finally, (7.45) for magnetosphere energy storage gives E_L = 2.04 $\times$ 10^{14} J. This is a relatively modest energy being less than 5% the yield of a single 1-megaton nuclear bomb (4.2 $\times$ 10^{15} J). Noting that the wire of the current loop is rather narrow compared with its diameter, inducing more mechanical rigidity in the wire could be achieved by forming it into a hollow tube of significantly greater radius. For example, a tube of 1 m radius would have a wall thickness of 0.334 mm. An ultimate limitation on increasing the diameter of such a tube and therefore thinning its walls comes from the fact that a thin surface layer of a superconductor (the London penetration depth – typically in the range 50–500 nm) is excluded from the superconducting behaviour. It cannot be afforded that too large a fraction of the cross-section is excluded from superconducting.

A check on the effectiveness of the magnetospheric drag needs to be performed. It is necessary to confirm that the interstellar protons impinging on the artificial magnetosphere at the maximum velocity of $c/2$ are sufficiently deflected by the magnetosphere that they will transfer a significant fraction of their momentum to the magnetosphere as has been assumed. It should be a sufficient criterion to show that the radius of curvature of these protons in the weakest part of the magnetospheric field at the magnetopause radius is smaller than the magnetopause radius. This should guarantee that the great majority of the protons are greatly deflected within the spatial scale of the magnetosphere. Calculating B_0 from (7.36) at 1.767 $\times$ 10^{14} T m^3, the magnetopause radius is defined by (7.29) as 3312 km. This is actually slightly smaller than r_{ring}, which means that the ring diameter has been pushed to its limit in this example, and the initial magnetosphere might be slightly annular. However, the magnetosphere expands significantly in size as the spacecraft decelerates.

The proton radius of curvature can be calculated by equating the force on the proton charge e at proton velocity v_p from the magnetic field strength B with the centripetal force, i.e. $Bev_p = mv_p^2/r$ or $r = mv_p/Be$. The magnetopause field can be estimated by using (7.27) and setting the radius as the magnetopause radius R_{MP}. Then, $v_p = c/2$ gives a proton radius of curvature of 322 km, which is less than a tenth of the magnetopause radius, thus confirming a satisfactory degree of proton deflection.

The journey phases and their distances and durations for the parameters adopted in the example case are summarized in Table 7.1. Thus, for example, the

Table 7.1   Journey phases and their distances and durations (X is an arbitrary distance in light years)

Journey phase	Distance (light years)	Duration (years)
Light sail laser boosting (1g)	0.119	0.5
Cruise	X	2X
Magnetospheric deceleration	0.754	7.8
Thermonuclear deceleration (1g)	0.00526	0.1
Total	0.87826 + X	8.4 + 2X

journey time to Tau Ceti, which lies 11.9 light years from Earth and is of a similar G-class star type to the Sun, is 30.5 years.

7.14   Targeting the light sail

In order to minimize the requisite power in the laser beam, it needs to be accurately targeted to fit the exact cross-section of the light sail. Since the angle subtended by the diameter of the light sail diminishes as it is accelerated away by the laser beam, the mirror directing the laser beam into the sail needs to have a continuously adaptable curvature to vary the focal distance. But ultimately, the angular width of the beam has a minimum angular cross-section at its focus determined by the diffraction of the beam. This diffraction angular width must fit the cross-section of the light sail at its narrowest angular width as viewed from the laser mirror, whilst it is operating. That occurs at its maximum boost distance.

The cruising speed has been defined as $c/2$ and the acceleration as $1g$, so the time to reach the cruising speed is $c/2g = 1.5 \times 10^7$ s or just under half a year. The distance travelled s in this half year of boosting is:

$$s = \frac{gt^2}{2} = 1.125 \times 10^{12} \text{km} = 7520 \text{ Astronomical Units (AU)} \qquad (7.50)$$

From (7.6), the angle subtended by the radius of the sail at this range is $r_{sail}/s = 8.3 \times 10^{-10}$ radians. If this corresponds roughly to the angle θ of the first minimum of the diffraction from the cross-section of the mirror and the wavelength is 500 nm, then, also from (7.6), the diameter d_A of the mirror must be greater than 801 m. In practice, however, in order to mitigate the power density on the mirror, given the vast power it must transmit, it will probably need to be at least hundreds of times wider than this.

There will additionally need to be a feedback system to keep the laser beam centered on the light sail, given the considerable pointing accuracy demanded by the tiny target angle. In short, this requires some kind of rim feature implemented on the sail that modifies the reflected light differently at different rim locations such that the reflected beam exhibits a detectable modification functionally dependent upon the direction of laser beam drift relative to the light sail. Something along the lines of rim strips that modify the wavelength of the light reflected from them should serve this purpose adequately. Clearly, the reflected color variation around the rim would have to be defined judiciously, and the sail would need to be kept in the same orientation so that the consequential tinting of the reflected beam would translate straightforwardly to a particular direction of drift of the beam pointing angle with respect to the sail. Alternatively, the vessel could use radio communications to relay beam incidence data back to base.

7.15   Early adoption within the solar system

Since laser-boosted light sails and magnetospheric ram pressure look attractive for solving the conundrums of human-lifetime interstellar travel, they could also offer good solutions for spacecraft operating within the solar system. When scaled down to meet the less stringent requirements of interplanetary transportation, these technologies may well grow rather than diminish in attractiveness. For example, accelerations much lower than $1g$ and peak velocities of just 1% of the speed of light would yield journey times of just days or weeks compared with the years or decades available using current chemical boosters. For such short journeys, much lighter spacecraft would be satisfactory for supporting human crew and passengers.

Furthermore, the ion densities in interplanetary space should be better than in interstellar space, notably due to the solar wind, which is a plasma with a number density between 3 and 10 particles per cubic centimeter near the Earth, decreasing as roughly the inverse square of distance from the Sun [19]. Since the solar wind also has a velocity ranging from 250 to 750 km/s directed somewhat outwards from the Sun, it is feasible in principle to use an artificial magnetosphere to sail this wind around the solar system. The pressure exerted by the solar wind is in the range of 1–6 nPa at the radius of the Earth's orbit from the Sun. This may be compared with the pressure of sunlight on a light sail at the same radius, which can be derived from the power density of solar radiation upon the Earth of $\sim$1.4 kW/m^2 using (7.7) [20]. This equates to a light pressure of 9333 nPa. This suggests that an artificial magnetosphere needs to be approaching a hundred times greater in diameter than a light sail to generate a comparable thrust. However, the superconducting current ring used to generate the magnetosphere need not be much greater in diameter than the equivalent solar sail and the ring could have practical advantages regarding mechanical stability and structural integrity. A limitation of artificial magnetospheres would be that they cannot exceed the speed of the solar wind particles and so they are limited to a few hundred km/s, but that is a million km/h, which might be regarded as adequate for interplanetary travel.

It would therefore seem likely that laser-boosted light sails and artificial magnetospheres will initially be developed to support transport systems designed to operate within the solar system. In this context, they may additionally serve as a testbed for the larger and higher specification systems required to effectuate interstellar travel within a human lifetime. It is perfectly practicable to envisage the development of these interplanetary transport systems on a timescale of a century or so from the present.

7.16   Conclusions

This chapter has specified a theoretical basis for travel to nearby star systems on a timescale well within the lifetimes of a human crew by means of electromagnetic propulsion technologies. A laser-boosted light sail accelerates the spacecraft at $1g$ for half a year, followed by a cruising phase at circa $c/2$. Deceleration is mainly effectuated by the drag of interstellar ions on an artificial magnetosphere induced around the spacecraft by a current in a superconducting ring. However, a final thermonuclear-powered phase of deceleration after the speed has been reduced below $c/10$ is necessary due to the progressive weakening of magnetospheric drag as the velocity relative to the interstellar ions reduces.

Nevertheless, it has also been shown that the engineering challenges in realizing this system of travel are immense. It implies a spacecraft having a mass of a million tonnes with a tantalum light sail just 22 nm thick formed from a single crystal of the metal. The laser required to feed this light sail might need to be tens of thousands of km long, and the adaptive laser mirror would need to be around 1000 km in diameter to handle the requisite power density. The power output from this laser would need to exceed a third of a Megaton per second continuously for 6 months. The superconducting ring used in the deceleration phase would have to be fabricated during the cruising phase by reforging the sail material. This ring would need to be around 10,000 km in diameter and maintain a fixed current of the order of 2 million Amps for up to 10 years.

These challenges are such that the timescale for them to be met probably extends to centuries. However, it has also been shown that both the laser boosting of light sails and the deployment of artificial magnetospheres are attractive means of interplanetary transportation. Therefore, it is possible to envisage their use within the solar system within the next century or so and that could be a steppingstone to scaled-up versions of the same technologies aimed at interstellar travel.

References

[1]   https://www.icc.dur.ac.uk/~tt/Lectures/Galaxies/LocalGroup/Back/12lys.html (accessed 10/8/2024)
[2]   https://en.wikipedia.org/wiki/Rocket_engine_nozzle (accessed 10/8/2024)
[3]   https://en.wikipedia.org/wiki/Breakthrough_Starshot (accessed 6/9/2024)
[4]   https://en.wikipedia.org/wiki/IKAROS (accessed 9/9/2024)
[5]   P. A. Tipler, *College Physics*. New York: Worth, 1987, p. 316
[6]   https://www.astro.princeton.edu/~gk/A403/hw1.pdf (accessed 18/08/2024)
[7]   S. N. Shore, IV.B The environment of the interstellar medium, in *Encyclopedia of Physical Science and Technology* (3rd edn), Amsterdam: Elsevier, 2003
[8]   https://www.sciencedirect.com/topics/physics-and-astronomy/interstellar-matter (accessed 28/08/2024)
[9]   https://www.princeton.edu/~maelabs/mae324/glos324/tantalum.htm (accessed 3/9/2024)
[10] https://physics.nist.gov/PhysRefData/Star/Text/PSTAR.html (accessed 3/9/2024)
[11] R. Dendy, *Plasma Physics: An Introductory Course*. Cambridge: Cambridge University Press, 1995. p. 234.
[12] https://en.wikipedia.org/wiki/Magnetopause (accessed 31/8/24)
[13] https://en.wikipedia.org/wiki/Conventional_superconductor (accessed 1/9/2024)

[14]   https://physicsworld.com/a/tackling-current-limits-in-superconductors/ (accessed 31/8/2024)

[15]   https://www.quora.com/Is-there-a-limit-to-how-much-energy-can-be-placed-into-a-superconductor (accessed 31/8/2024)

[16]   https://en.wikipedia.org/wiki/Superconducting_wire (accessed 31/8/2024)

[17]   https://learnemc.com/ext/calculators/inductance/circle.html   (accessed   2/9/2024)

[18]   https://periodictable.com/Elements/073/data.html (accessed 3/9/2024)

[19]   https://en.wikipedia.org/wiki/Solar_wind (accessed2/9/2024)

[20]   R. Rosner, *Macmillan Encyclopedia of Physics*, vol. 4. New York: Simon & Schuster, 1996, p. 1545.

Chapter 8

The future evolution of liquid crystal devices and photonics

Garry Lester[1]

The subject of photonics is vast, encompassing systems within the wavelength range not just of visible light but extending from virtually millimetric terahertz wavelengths to sub-micron ultraviolet wavelengths. Across this spectrum, it also covers technologies for generating, directing, controlling and detecting these electromagnetic waves. Optoelectronics can be regarded as a subset of photonics, relating to the devices and interactions used to interface between electronics and optics. It is therefore not possible to encompass all aspects of optoelectronics and photonics in a single chapter or even a single volume.

Instead, this chapter will focus mainly on liquid crystal device (LCD) technologies. This is partly because of the major contribution LCDs have made in establishing and developing information display applications. It is also because LCDs offer a wide range of optical switching effects over an ever-expanding range of wavelengths. They offer much more than passive display capabilities. LCDs can contribute electronic reconfigurability to an otherwise fixed-function optical system.

8.1 Photonics in society

Optics and photonics has in relatively recent decades become an established and essential part of the way modern technological society functions. In key areas such as displays, communications and sensing and imaging, photonic technologies are now very much an integral part of daily working and social lives. Flat panel display technologies are an essential part of information and entertainment systems, having almost entirely replaced cathode ray tubes and mechanical displays. Optical fiber communications support the transmission of large volumes of information to meet the high-resolution display needs of modern society. From transmission over metre distances in the 1960s through improvements in optically transparent materials and signal amplification, it became feasible to transmit data optically over hundreds or thousands of kilometres. The ubiquitous optical mouse functioning as a reliable

[1]EMC Department, MBDA, UK (formerly L-electronics Ltd, UK and Department of Engineering, University of Exeter, UK)

computer interface pointing device uses oblique illumination and electronic image correlation to detect motion relative to the surface it rests upon. Optical coherence tomography to image the eye retina is now routinely available on the high street for monitoring and diagnostic visualisation. Also, less immediately obvious but no less important is the use of LiDAR and similar optical spectroscopy techniques in applications such as environmental sensing.

8.2   Liquid crystal technologies

Building on this established base of the demonstrated importance of optical and photonic technologies, research is ongoing to further develop these fields and to explore new techniques, materials and physical effects that might be exploited. Liquid crystal technologies are now discussed in some detail for their historical significance, the interesting electromagnetic aspects of both the low-frequency dielectric coupling and the optical frequency effects and, finally, for the transfer of these technologies into other research and application areas.

8.2.1   Key liquid crystal properties

There are two groups of liquid crystal materials: thermotropic liquid crystals driven by temperature and lyotropic liquid crystals driven by solvent concentration. Lyotropic liquid crystal phases exist that depend upon concentration within a solvent, but the liquid crystals considered here are thermotropic liquid crystals where the phase structure is a function of temperature. Lyotropic systems form structures that are in some ways analogous to thermotropic structures and are of fundamental importance to many biological systems and surfactants.

Thermotropic liquid crystals are a group of phases of matter exhibited by molecules that have some significant physical anisotropy, typically with long thin molecular structures. On heating from the solid phase, a material with one or more liquid crystal phases will undergo phase transitions through states that have both some degree of orientational order in common with the solid phase, but also fluidity in that the molecules can move relative to each other and exchange places in common with the liquid phase. These are the liquid crystal phases, and on further heating, the material will become an isotropic liquid with no overall orientational ordering of the molecules. The phase sequence is shown in Figure 8.1. In the bulk liquid crystal phases, the local mean molecular alignment direction is denoted by the director. A more detailed discussion is given in [1].

The anisotropy of the liquid crystal molecules gives rise to different electron mobilities in different directions with respect to the molecular axis. This manifests itself in the bulk material as an anisotropy in the dielectric constant depending on the direction of the applied electric field relative to the director. At higher optical frequencies, this becomes an anisotropic refractive index or birefringence, the refractive index depending on the polarisation of the illumination relative to the director. These anisotropies are particularly useful in the case of liquid crystals as

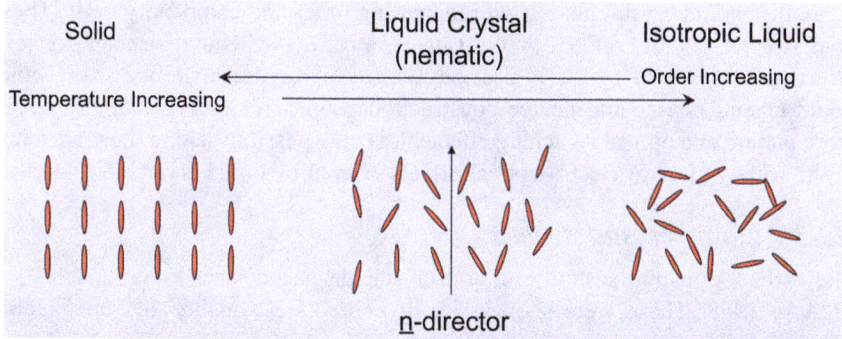

Figure 8.1   Solid-to-liquid phase transition sequence through liquid crystal phases

they give both a useful optical effect and a mechanism for control of the orientation of the optical effect through the dielectric anisotropy coupling with an externally applied electric field. Most liquid crystal technologies exploit the low-frequency coupling between the dielectric anisotropy of liquid crystal molecules and an applied low-frequency electric field to provide a handle to induce reorientation of the liquid crystal molecules.

In addition, the physical anisotropy will give a degree of cooperative ordering of the molecules against an anisotropic treated surface. In many devices, this gives the field off-state alignment of the liquid crystal. Importantly, these switching effects occur with fields of the order of 10^5 V/m, and in visible wavelength devices, a liquid crystal layer of only a few micrometres is required for a useful optical effect, so the applied voltage required is typically only a few volts. This makes the simpler devices useful in battery-powered applications.

Fundamental to LCDs is that they do not generate light but rather control light. This makes them ideal for optical devices where the optical transfer function is to be reconfigured by the application of an electric field. The polarisation dependence due to birefringence can be a limitation in some device geometries if it is required to work with randomly polarised light such as sunlight.

8.2.2   Liquid crystal phases

The nematic phase is the 'simplest' thermotropic liquid crystal phase, with the liquid crystal molecules having orientational but not positional order. The molecules tend to align giving orientational order, but they are also able to move relative to each other. It is this combination of physical ordering of alignment combined with fluidity that is exploited in most LCDs.

As the liquid crystal molecules are in constant thermal motion, their relative orientations will fluctuate around the average alignment direction, the director. The degree of fluctuation from the average alignment direction is represented by the order parameter. In a liquid crystal, the director may vary naturally through the volume of the material, often forming intricate patterns and structures, but in devices, it is usually desirable to have a controlled, in most cases uniform, director alignment.

Other liquid crystal phases with increasing order and complexity exist. These give rise to a wide variety of optically controlled switching mechanisms and effects. The majority of LCDs use the nematic phase, but there are many other liquid crystal phases and device construction geometries with potentially useful reorientation and optical switching effects. It is only possible here to illustrate a few of the wide variety of electro-optical effects offered by liquid crystal technology.

8.2.3   Liquid crystal devices

The earliest dynamic scattering liquid crystal display devices were launched by RCA in 1968. These were short-lived, of poor optical quality and not notably successful. They did however demonstrate a need and potential market for an electronic display device.

The twisted nematic (TN) LCDs [2] were not the only available flat panel display technology. However, at the time, they offered significant advantages. LCDs became ubiquitous as low-power, low-cost displays for calculators, watches, instrumentation and similar devices. One of the notable features of the technology has been its scalability, with progressive investment levels from these early simple displays through laptops and similar screens to large area complex display panels. The requirements of the display industry have driven significant investment in LCD materials, surface alignment coatings, fabrication processes and optical effects that are now enablers for the development of LCDs for other applications.

8.2.4   Simplest LCD

Probably the simplest LCD has the liquid crystal material contained between two transparent substrates, as shown in Figure 8.2. This serves as a useful introduction as it shares many of its key attributes with more complex LCDs.

To construct a useful device, it is necessary to impart a uniform director alignment on the nematic liquid crystal; this is most commonly achieved using surface treatments on the substrates. The surface treatments are given a preferred surface alignment direction by direct rubbing, by oblique evaporation of materials onto the surface or by photolithographic patterning, denoted S.A. in the figures.

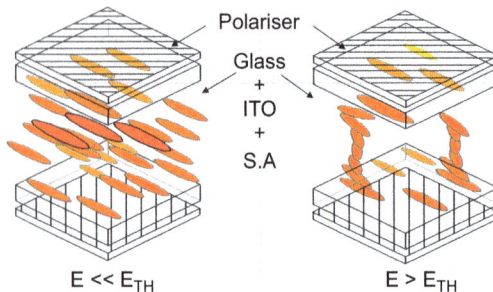

Figure 8.2   Variable birefringence device

This physical anisotropy of the surface couples to the liquid crystal director at the surface, and this propagates some way into the bulk liquid crystal material. For sufficiently closely spaced substrates, this alignment will propagate through the whole device. Typically, this surface-induced alignment will propagate up to a few tens of micrometres into the material before being disrupted by thermal effects. With parallel rubbing directions for the surface treatments, this will give a uniformly aligned 'slab' of liquid crystal with the director along the rubbing direction of the surface treatments. The optic axis is the direction of propagation for which there is no difference in refractive index for any polarisation. For liquid crystal materials, this is normally approximately along the liquid crystal long axis. Optically, this will behave as any other birefringent crystal with wavelength-dependent phase shifts between the electric field components of the incident electromagnetic wave aligned with or perpendicular to the optic axis. Incident polarised light may be regarded as having two orthogonal components. The retardation of these two polarisation components relative to each other depends on the difference between the refractive indices Δn and the thickness of the birefringent layer d.

The surface treatments on the inner faces of the substrates will control the initial liquid crystal alignment, but in order to perform a useful optical function, it is necessary to modify the liquid crystal orientation. This is achieved through the application of a low frequency (typically a few kilohertz) electric field coupling to the dielectric anisotropy. In order to apply this field, electrodes are required, and for most applications, these must also be transparent at the wavelength of operation. Indium tin oxide (ITO) coatings are widely used as electrodes as they are transparent at visible optical wavelengths and provide surface resistivities of tens of Ohms per square. As the liquid crystal has a high resistivity and responds to field rather than conduction, these electrode resistivities are sufficiently low to drive LCDs efficiently. Beneficially, as the current required to drive the reorientation is very low, these devices consume very little power.

Due to the liquid crystal birefringence, this device will inherently have a strong wavelength, polarisation and viewing angle dependence. Incident linearly polarised light may be described as having electric field components both along and perpendicular to the liquid crystal optic axis. Each of these components will experience a different refractive index on passing through the device. Depending upon the incident polarisation, device thickness and refractive indices, the resulting polarisation on having passed through the device will be complex.

For the simple case of normally incident illumination linearly polarised at 45° to the optic axis, the two components will see a relative retardation of $2\pi.\Delta n.d/\lambda$ radians, where Δn is the difference between the two refractive indices, d is the thickness of the birefringent layer and λ is the wavelength. For a given fixed thickness d and birefringence Δn, there will be a strong wavelength dependence when the device is placed between crossed polarisers, as shown in Figure 8.3 for $d = 10$ μm and $Dn = 0.225$. This figure is somewhat idealised in that the wavelength-dependent effects of the polarisers and other layers within the device have been ignored.

An electric field above the threshold field applied to the liquid crystal (LC) layer via transparent electrodes will couple into the low-frequency dielectric anisotropy $\Delta\varepsilon$ and

Transmission vs Wavelength for Δnd = 0.225×10 μm

$$I_t = I_o \sin 2 \left(\frac{\pi \cdot \Delta n d}{\lambda} \right)$$

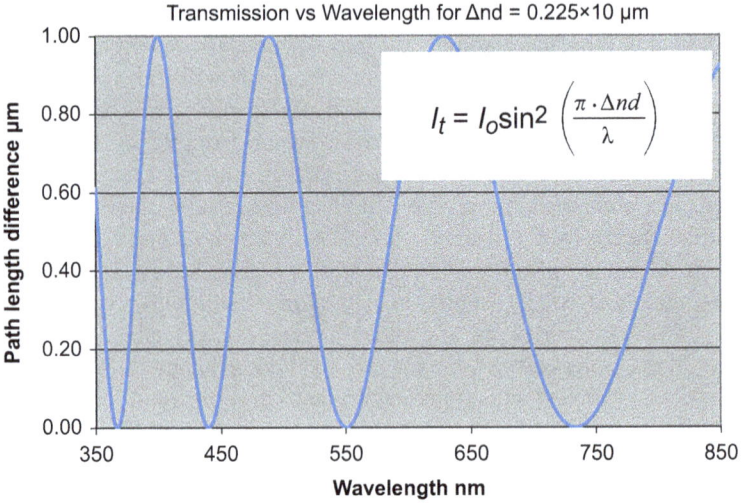

Figure 8.3   Idealised transmission of a birefringent device

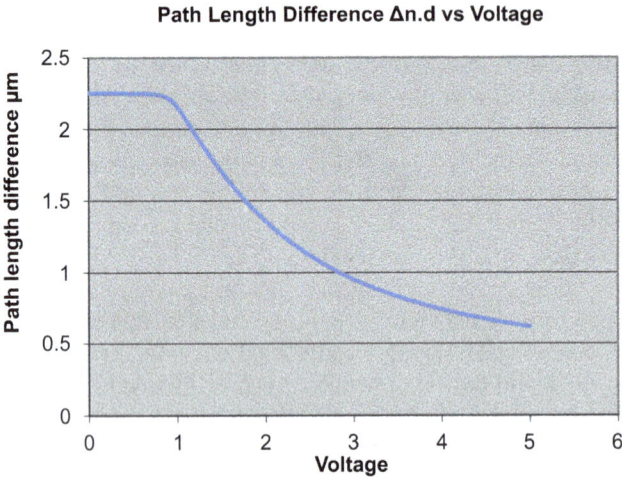

Path Length Difference Δn.d vs Voltage

Figure 8.4   Path length variation with applied voltage

cause reorientation of the liquid crystal director. Figure 8.4 shows the change of retardation for a 10 μm liquid crystal layer of $Dn = 0.225$ with applied voltage. For propagation of light normal to the substrates, this will reduce the effective birefringence Δn and move the transmission peaks to shorter and shorter wavelengths. Ultimately, with minimal remaining birefringence, the device will appear non-transmitting or black.

Even this simple device offers a potentially useful switchable optical component as an electrically controlled waveplate. As a display device, this might offer simple device construction where only a limited range of colours is required. For a black-white display or display with accurate colour rendition achieved through filtering of a basically black-white display, the inherent colouration and variation of colour with the viewing angle of this device is a severe limitation.

This simple device does illustrate one of the key features of LCDs when compared to other photonic technologies. LCDs control the propagation of light through the refractive index profiles generated either naturally as part of the liquid crystal phase or by the applied field structure. Many other photonics technologies only generate light. LC technologies will often require a light source for display applications, but the ability to control light also opens up other applications in photonics.

8.2.5   Dual frequency and negative delta epsilon materials

For many LCD materials, the optic axis aligns with the applied field as this aligns with the dielectric anisotropy. There are some materials in which the opposite is true such that the liquid crystal will align perpendicular to the applied electric field. There are also some materials where the sign of the dielectric anisotropy depends upon the frequency of the applied field, exhibiting both positive and negative dielectric anisotropy behaviour depending on the frequency.

8.2.6   LCDs for visual displays

One of the largest applications of thermotropic LCDs is in the display industry. This has attracted a significant investment in materials, processing techniques and technologies that almost certainly would not otherwise have existed. The development of liquid crystal displays has also benefitted from the advantage of scalability. The demonstrable and profitable markets for display devices of increasing size have taken liquid crystal technology from monochrome display devices of a few square centimetres to display panels of more than 1 m^2 with a wide gamut of colours. This has necessitated investments into materials both for substrates and for liquid crystals, large substrate handling techniques and surface treatment technologies.

It is not reasonable to mention liquid display technology without at least some discussion of the twisted nematic (TN) device. For display applications, the TN device probably still is, at least in terms of a number of devices if not display area, the most prevalent display technology. The TN device is the ubiquitous display for small area, seven-segment (character definition) applications such as watches, calculators, multi-meters and so on. The TN device shares many of the construction features of the simple variable birefringence device, with transparent substrates, ITO electrodes and surface treatment to control LC alignment. However, in terms of the liquid crystal alignment and the optical behaviour, it differs in that the alignment of the liquid crystals at the top and bottom surfaces are at 90°. This, along with some degree of doping of the liquid crystal with a chiral twisting agent, supports a field-off helical structure within the liquid crystal. Optically, this is complex to analyse, but provided some conditions of wavelength relative to cell thickness and birefringence are met [3], the helical birefringent structure rotates the plane of incoming linear polarised light through 90°. This waveguiding polarisation rotation may be engineered to be relatively broadband across the visible spectrum giving an essentially white device. The application of an electric field will reorient the molecular director perpendicular to the substrates, destroying the helix and preventing the polarisation rotation. With appropriately patterned electrodes and fitted with polarisers oriented along the surface alignment directions, this gives good quality black figures on a white background display device. The TN device is of good optical quality and very low power consumption, the optical effect being dependent upon the applied voltage with minimal current consumption, and is straightforward to manufacture in significant numbers. TN devices are imperfect in many ways, but in appropriate applications demonstrate the efficacy of electronically driven display technologies. Thus they paved the way for the modern displays market.

The TN device in its purest form does not offer colour, has a restricted viewing angle and will not readily form a complex matrix display with high contrast. The early promise of 'hang on the wall' TV-like displays took significant further innovation and investment over some decades to realise. With the addition of colour filters, active pixel driving systems, novel LC device geometries and back-lighting systems, high-quality flat panel displays are now ubiquitous.

8.2.7 Matrix addressing schemes

The devices as described so far only provide a uniform optical effect across the whole device. For most applications, some spatial variation of the optical effect is required. This is achieved through patterning of the electrodes such that the field is only applied locally where LC reorientation is required.

Depending on the complexity of the spatial field pattern required, for a small number of electrodes, it may be possible to bring the connections out to the edge of the device, facilitating direct driving of the electrode elements. For more complex spatial field variations, a matrix and multiplexed addressing scheme may be required.

In such schemes, the pixels are driven with a data voltage on rows that combines with a bias voltage on columns to give the required driving voltage at the individual pixels. For a simple passive matrix array with N rows and N columns, as the number of lines increases, the RMS voltage applied to any one pixel tends towards the average mid-bias voltage as only $1/N$th of the time is spent specifically addressing any one line. The rest of the time the pixel will be exposed to part of the waveforms addressing the other lines. The difference between the voltage when selected to be on or to be off is what drives the optical difference between pixels; with more rows, this voltage difference becomes ever smaller.

Nematic liquid crystal materials respond and reorient to the RMS-applied electric field. For a useable optical difference with a large number of rows and columns, this requires the liquid crystal material to have a very sharp transition in the molecular reorientation with respect to the change in driving voltage. This leads to optical compromises in the material design and the number of addressable lines is relatively limited, a few hundred at best. One way of achieving this steep transition is the supertwist nematic materials and devices [4]. In these, the construction is similar to the TN device except that the liquid crystal goes through a rotation of greater than 90° through the cell. The optics are more complex, with residual birefringence, yet with compensation films, it is possible to make a display of good optical quality.

Active matrix addressing schemes largely avoid the requirement for a steep transition but are considerably more complex, and the active electronic devices required to be fabricated within each pixel area significantly degrade the optical transmission. For redundancy, two or more transistors are used on each pixel.

In order to achieve high degrees of multiplexing, it is now established technology to incorporate active electronics at the pixel level; this allows the liquid crystal material to be locally optimised for its optical properties. With the addition of optical colour filters and one or more transistors on each pixel, such displays are far more complex than the simple TN-style devices.

Liquid crystal on silicon devices extend the active matrix device to the point where the active matrix becomes the substrate. The liquid crystal layer sits on top of a silicon wafer containing the driving circuitry with the device being used in reflection.

8.2.8   In-plane switching and vertically aligned nematic (VAN) device configurations

The device construction described so far applies the electric field perpendicular to the substrates with the consequent reorientation of the liquid crystal towards the substrate normal. If electrodes are arranged to give a field in the plane of the substrates, then the liquid crystal will of course align with this field and reorient in the plane of the substrates. With no field applied, the liquid crystal molecules align along the surface treatment direction. The electric field in the plane of the substrates is generated by adjacent electrodes on the same substrate but with opposite polarity. The field between these electrodes has a significant component in the plane of the substrates that reorients the molecules closest to the substrate, and this reorientation then propagates through the bulk of the liquid crystal. This technique is now commonly applied in display devices as it gives a better visual optical effect over a wider range of viewing angles.

Another configuration frequently used in display devices is the vertically aligned nematic (VA or VAN) device. In this configuration, the liquid crystal with field off is initially aligned perpendicular to the substrates. The liquid crystal materials used in these devices have negative delta epsilon, so with a field applied, the molecules move away from the perpendicular, inducing birefringence and allowing light to be transmitted. The vertically aligned state is able to give very low light transmission and therefore high contrast between light and dark states. The terms tip and tilt are avoided here as they are not used consistently between researchers working on different styles of device, potentially causing confusion.

8.2.9   Chirality and chiral liquid crystal phases

If the molecules of the liquid crystal are handed, such that a mirror reflection of the molecule cannot be overlaid on the original, and the liquid crystal material contains more of one handedness than the other, then the liquid crystal phases are likely to form twisted structures. In the chiral nematic phase, for example, the molecules cooperatively align but not with the molecular axes exactly parallel. This gives rise to a helical pattern in the alignment with the pitch dependent upon the molecular structure, temperature and applied field. In fact, it is normal practice for the liquid crystal material within a TN device to have some degree of chirality in order to bias the twist slightly to the left or right-handed state; otherwise, any ambiguity can give rise to regions of different-handed twists and defect lines between them, spoiling the uniformity of the device.

The chiral nematic phase is often also referred to as the cholesteric phase. The helical structure in this phase gives rise to a number of useful optical effects. The birefringent helix provides for wavelength-selective Bragg reflections, with the wavelength dependent upon the pitch of the helix. The pitch is extremely temperature dependent and forms the basis of sensitive passive liquid crystal thermometers [5].

With the pitch of the liquid crystal phase shorter than the physical dimensions of the device and the alignment not controlled to be uniformly parallel to the substrates, a focal conic texture can form. In this molecular arrangement, regions of varying liquid crystal orientation form, still with helices within the different regions. At the boundaries between these regions, the differences in refractive index give rise to a strong scattering effect and a cloudy appearance at visible light wavelengths.

8.2.10   Smectic and ferroelectric devices

Some materials exhibit smectic phases between the nematic phase and the solid phase. In the smectic phases there is not just orientational order but some degree of positional order. There are several smectic phases with varying degrees of positional order. In the smectic A phase, there is a tendency for the molecules to lie parallel to each other in what is often referred to as layers but are actually planes of increased density. In this phase, the molecular axis is normal to the layers. It is important to remember that these materials are still fluid albeit significantly more viscous than the nematic phase. Coupling of an electric field to the dielectric anisotropy will give rise to a molecular reorientation in the smectic A phase. In the smectic C phase, there is a tendency for the molecular axis to tilt away from the layer normal by an angle that depends on both the material and temperature, and in the chiral version of the smectic C phase, the chirality manifests itself as a precession of this tilt around the layer normal.

Constraining the smectic phase with closely spaced surface-treated substrates such that this helix is unwrapped gives rise to a spontaneous polarisation and ferroelectric switching behaviour [6]. The coupling of an electric field to this polarisation is a switching mechanism in addition to the dielectric anisotropy coupling. The application of an electric field between the substrates couples with this polarisation and causes molecular rotation around a cone surface to take up the same tilt angle from the layer normal but on the opposite side of the cone. This gives a rotation of the birefringent optic axis in the plane of the substrates of twice

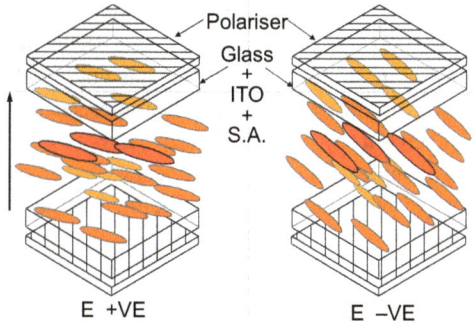

Figure 8.5   Ferroelectric chiral smectic C device

the tilt angle (Figure 8.5). This can take place in the order of a few microseconds. Optically this device would typically be configured to have a retardation $\Delta n.d$ of $\lambda/2$ with a liquid crystal layer thickness of only around 1–2 μm. Viewed through crossed polarisers with one polariser along and the other perpendicular to the optic axis, there is no phase shift of the polarisation components, and the light is blocked. If the liquid crystal is optimised to rotate the optic axis through 45°, this imparts a $\lambda/2$ phase shift of one polarisation component, effectively rotating the polarisation through 90° so that light is then transmitted.

The switching of these devices is in the order of microseconds and can be made bistable in that with the field removed, the molecular axis retains its position. Bistability has significant benefits in addressing in that displays with large numbers of rows can be addressed with a simple electrode structure similar to that for passive matrix addressing. Also, if the stable state is maintained with no applied field, then the device will consume minimal power in addressing the liquid crystal. This was one of the motivations for attempts to commercialise flat panel displays using ferro-electric liquid crystal technology. Some manufacturing problems were encountered for this type of device as the liquid crystal layer needed to be only around 1 or 2 μm thick to reliably align the liquid crystal, but the main problem was that this alignment was not stable in use. Deformation of the flat panel glass substrates could introduce defects into the alignment that, unlike most nematic devices, would not recover. For smaller more rigid devices, this is not such a problem.

8.2.11   Blue phases

Generally, it might be expected that liquid crystals at higher temperatures become less viscous with consequently faster response times. The blue phase is a chiral structure that forms close to the transition to an isotropic liquid. This is a complex structure of cylindrical helices. The chiral structure gives rise to wavelength-dependent optical properties. Although blue phases can have selective reflection at a range of wavelengths, when these were first observed, the predominant colour of reflection was blue, hence the name of the phase. Blue phase devices do exhibit rapid switching, demonstrably of the order of 100 μs, but only work effectively over a limited temperature range. A commercial blue phase liquid crystal TV was demonstrated by Samsung in 2008; however, this has not yet at the time of writing materialised in the form of a widely available commercial product. This is reminiscent of the announcement by Canon in the early 1990s of a commercially available display based on ferroelectric liquid crystal technology. Only a small number of units were ever produced before being overtaken by active-matrix devices.

8.2.12   Optically scattering and other devices

Liquid crystals offer optical effects other than birefringence devices. Wherever there are refractive index changes these will change the path of light passing through the material. With many closely spaced changes of refractive index, as there are with suspended water droplets in fog, the light is scattered. A similar electrically selected optical scattering effect is achieved in some LCDs through a range of mechanisms, but all exploiting the field-induced change of the spatial refractive index distribution that is possible with liquid crystals.

In the polymer dispersed liquid crystal (PDLC) device shown in Figure 8.6, droplets of liquid crystal are suspended in a solid polymer matrix. Due to surface interactions at the liquid crystal polymer interface, the liquid crystal within the droplets will take up radial or planar patterns depending upon the materials. With no field applied, the orientation of these patterns within the droplets is random with many mismatches of refractive index and consequent strong optical scattering, giving a milky appearance. With a field applied the liquid crystal will reorient and, with correctly matched liquid crystal and polymer refractive indices, will significantly reduce the scattering effect. This optical effect has been applied in switchable windows and office privacy screens but can also give high contrast ratios in optical instrumentation systems, where even small angles of scattering away from the optic axis no longer form part of the measurement.

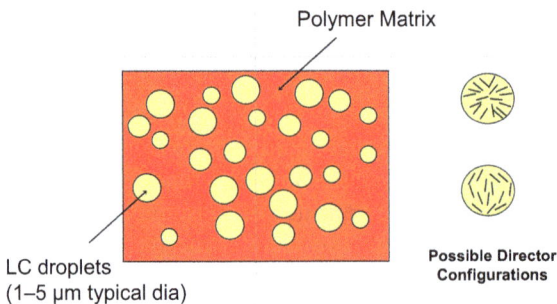

Figure 8.6   Polymer dispersed liquid crystal scattering device geometry

Optically similar scattering effects can be achieved by inducing liquid crystal textures, where there are small regions of co-aligned liquid crystal, but overall, the orientations of these domains are randomly distributed. This occurs, for example, in the chiral nematic or cholesteric phase, where the application of fields of different amplitudes and frequencies can be used to switch between uniform alignment or alignment broken up into sub-domains [7]. Some of these switching mechanisms offer two or more distinct stable states with zero field applied. Stable-switched states have significant advantages in that they do not require power once switched into the required state.

8.3 Liquid crystal non-display devices

The device geometries discussed so far have been developed mainly for display applications. The ability of LCD effects to influence light propagation also opens up non-display applications for reconfiguration of optical systems. The control of optical functions with liquid crystal technologies accesses applications in instrumentation, image processing and sensing through modulation of the optical function under electronic control. The possible device geometries are many and varied and selected solutions are likely to depend upon application as there is no one ideal technology. At the simplest level, the on–off switching of devices may be used to select alternative pathways through an optical system. More generally, the devices can exploit control of the spatial variation of refractive index or birefringence controlled through an applied electric field to make patterns that constitute, for example, lenses or holographic components [8,9].

8.3.1 Patterned electrode devices

One approach to inducing a refractive index profile is to use specifically shaped electrode patterns. Where there are electrodes, the liquid crystal will reorient locally, changing the optical effect. Matrix addressing is an established technique from the displays industry that has been used to produce reconfigurable optical devices. Such devices have been used to demonstrate switchable diffracting devices such as gratings and holograms. However, the regions between electrodes that either do not form part of the desired refractive index pattern or are black-masked-off give rise to a periodic pattern within any optical pattern written to the device. This regular grid in turn gives rise to diffraction artefacts in the optical function.

For fixed-function switchable devices, it is advantageous to directly pattern the required profile into the electrodes. This has been used to fabricate switchable Fresnel and Gabor lenses [10] and would be effective for a wide range of binary diffractive elements. In order to maintain connectedness of the electrodes, it would be necessary to include the electrode connectivity as part of the cost functions used to design the diffractive element.

8.3.2 Optical elements modulated by a liquid crystal

An alternative to inducing a refractive index pattern using electrodes is to create a diffractive element in a transparent structure and combine this with a liquid crystal,

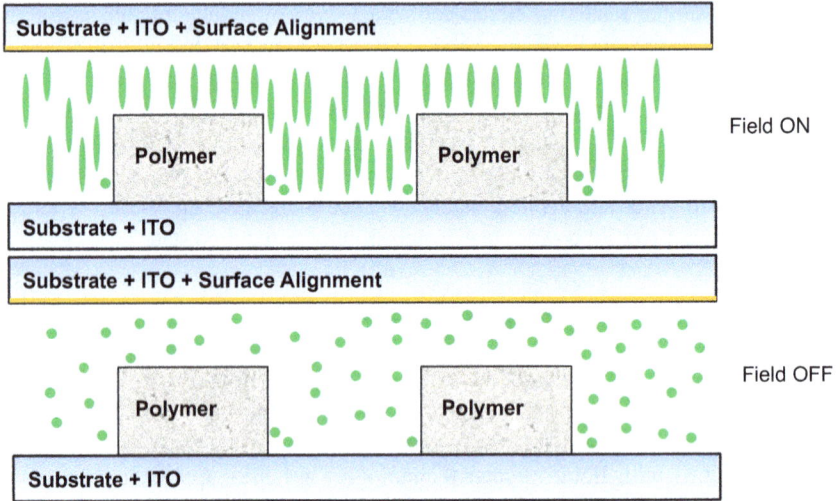

Figure 8.7   Polymer structure LCD

as shown in Figure 8.7. In a similar way to the PDLC devices, the liquid crystal provides a mechanism for an electric field to induce a change of refractive index match between the polymer and the liquid crystal. This modulates the optical path length difference between polymer and liquid crystal regions, allowing the diffractive optical effect to be switched on/off.

This type of technology enables the fabrication of switchable holograms or other optical devices and the device construction allows for easily replicated optical functions to be electrically modulated. The structures might be replicated using embossing or a similar fabrication technique. As binary or multi-level phase devices, these can be designed to be optically efficient with low insertion loss.

8.3.3   Optically switched optical devices

For all optical systems needing to avoid transferring signals between the electronic and photonic regimes, optically controlled optical functions are required. The low power drive requirement of LC devices allows them to be optically addressed through the incorporation of a photoconducting layer and application of a continuous bias voltage. Figure 8.8 shows an optically addressed spatial light modulator (OASLM). Illumination of the photoconductor with spatially varying intensity will optically enable only the illuminated areas of the optically controlled device to reorient [11]. When illuminated by another wavelength, this pattern can then form the input to the next stage of the optical system.

Directly optically responsive materials offer an alternative route to optically addressed optical devices. Liquid crystal phases are very sensitive to changes in the shape or conformation of the molecular constituents. Materials have been demonstrated that exhibit a reversible change of phase under illumination that might be used in an optically activated optical device [12].

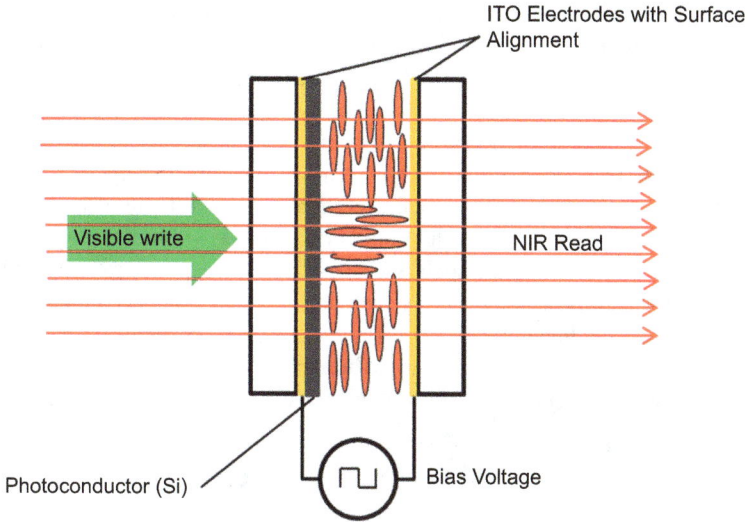

ITO Electrodes with Surface Alignment

Visible write

NIR Read

Photoconductor (Si)

Bias Voltage

Figure 8.8   Optically controlled device geometry

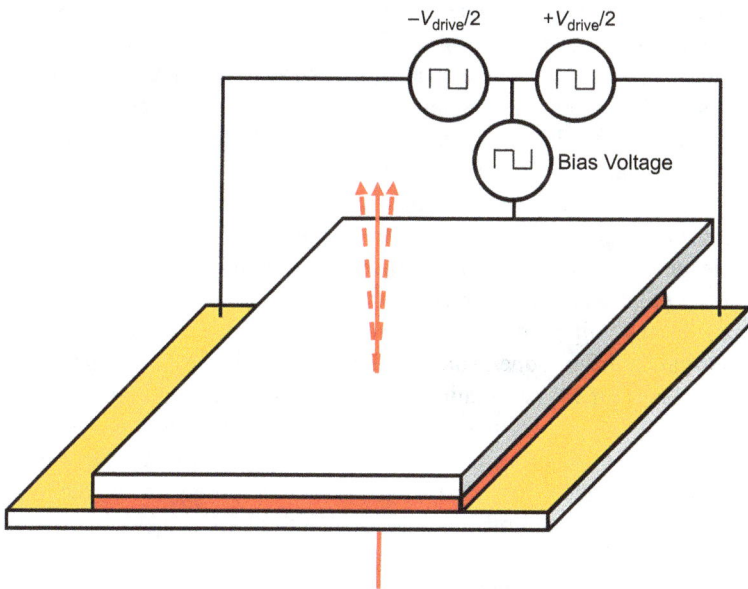

$-V_{drive}/2$ $+V_{drive}/2$

Bias Voltage

Figure 8.9   Beam deflector device construction and driving

8.3.4   Prismatic beam steerers

Patterned electrodes allow different field strengths to be selectively applied to different regions of the LCD. It is difficult with separate disconnected electrodes to apply a continuously varying field across the device without creating artefacts due to the fringing fields between the electrodes. Figure 8.9 shows a device construction and driving scheme with a continuous electrode driven differentially to produce a continuously varying field across the device. The reorientation of liquid crystal molecules will follow this field and give a continuously varying refractive index across the device for one polarisation. The variation of refractive index optically forms a prism and this can be adjusted continuously in time by varying the applied field. The maximum optical effect that can be achieved depends upon the birefringence of the liquid crystal and the thickness of the liquid crystal layer within the device.

Due to the need in display applications to have a steep electro-optic switching characteristic for multiplexed addressing, such liquid crystal materials are readily available and require only hundreds of millivolts difference to cover the whole range of molecular reorientation. A bias voltage would be applied across the whole of the electrode area with a potential difference across the device to create the required prismatic effect.

8.4   Future liquid crystal electro-optic technologies

8.4.1   Technologies currently in research and development

8.4.1.1   Flat panel displays

Display technologies will continue to be ever more important in our society and therefore a significant part of the photonics industry. While flat panel displays will continue to evolve, the more significant developments in liquid crystal technology are likely to be within alternative display technologies and non-display devices. These developments are particularly well facilitated by the ability of LC devices to control the propagation of light, so that complex optical functions may be achieved and controlled.

Flat panel LC display technologies have achieved high resolution, excellent colour gamut, high contrast over wide viewing angles and large areas. The achievable pixel resolution, colour gamut and viewing angles are now arguably reaching the point where the perceived improvements in display performance for most applications will be only slight. Although it would be easy to dismiss LC displays as having been 'done', some applications need extremely high resolutions, not just spatially but in grey scaling or colour rendition. For example, the reproduction of X-ray images requires an accurate rendition of grey levels in order to show soft tissue details. Dynamic range, especially the dark state, can be problematic for applications such as simulation of night conditions in simulator displays where even a small amount of light leakage can limit the achievable effect.

8.4.1.2   Other display technologies

Although there is now a wealth of flat panel displays available, there are also ever-growing markets in other types of display devices. These include, for example, head-up displays (HUDs) for all manner of vehicle types and directly worn near eye devices.

For near-eye displays, there are numerous configurations. Some project an entirely artificial image to the eye or eyes, while others augment a scene with superimposed supplementary information. Photonic display technologies are well suited to generating near-eye images or overwriting information onto a real-world scene, although in both approaches the optics required to correctly place the image size and plane with respect to the eye are complex.

Although, at first glance, it would appear fundamentally simple and straightforward to supply a stereoscopic pair of images to the eyes to give depth perception, this is of limited visual value and can lead to severe discomfort and eye strain for many users. Many problems arise due to the disparity between the focal plane and the left and right gaze point angles, the vergence-accommodation conflict (VAC), when the eyes are focussed at a different distance than the convergence point of the gaze-vectors from both eyes. The controlled optical devices offered by liquid crystal and now also other fluidic technologies provide for control of focus, although, for successful use in most applications, tracking of eye gaze point within the image and extraction of the corresponding depth information is required [13]. For artificially generated scenes, the three-dimensional (3D) information is readily available, whereas for the display of natural scenes, depth information extraction is required, and that is not readily available in a single camera image.

For effective accommodation correction, only a relatively small number of focal lengths are actually required rather than continuous focus, but focal switching must be comparable to accommodation timescales. A rapidly switching multi-focus lens system has been demonstrated using polarisation rotation with birefringent optics [14]. Something similar in combination with gaze tracking might be used for accommodation compensation. Low-cost eye gaze tracking is now available for gaming systems and extraction of depth information from stereoscopic image information is well developed, but the technology requires the combination of gaze tracking and depth data extraction at video frame rates with minimal time lag. It is likely that research in this area will bring to light various visual artefacts that will require attention.

For glasses-free viewing, flat panel lenticular displays with an opaque screen to restrict views of the display from either eye, as in Figure 8.10, are possible, and

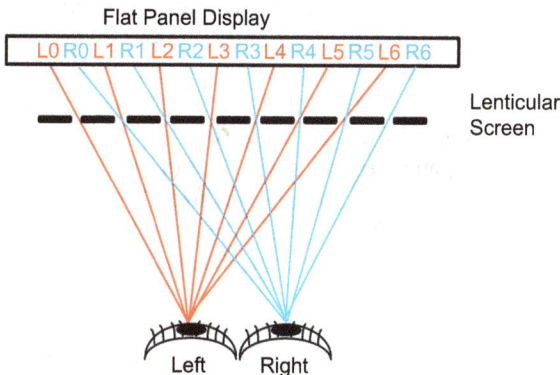

Figure 8.10   Lenticular stereoscopic display geometry

micro-LEDs or micro-pixel LCD panels are effective for this application. However, being essentially stereoscopic, these types of displays suffer from similar disparity effects and viewing angle ambiguity. They also sacrifice resolution in favour of realising the multiple views required. Furthermore, having a fixed focal plane, they do not offer the accommodation adjustment of head-worn devices, which incorporate variable focus lenses.

8.4.2 Materials and switching effects across the spectrum

8.4.2.1 Visible and near visible LC devices

Investment in the displays industry has provided a wide range of materials, driving techniques, and technologies. Much of this technology is optimised for operation at visible wavelengths. However, there are many potential applications involving reconfiguration and control of optical systems at other wavelengths.

The displays market and applications have driven research into liquid crystal materials such that there are materials available with liquid crystal phases over wide temperature ranges. Liquid crystal parameters do vary significantly with temperature, so that, typically, driving voltages and waveforms may need to be adjusted for optimum performance. Where operation outside of these ranges is required, it is possible to temperature control the device environment in much the same way as for crystal oscillators or similar devices.

8.4.2.2 Devices operating at wavelengths beyond the visible range

Materials developed and optimised for display applications are not necessarily optimised for use beyond visible wavelengths or for other types of applications. However, liquid crystal materials and devices are currently under active research and development across a wide spectral range. These devices might be used in optical instrumentation or imaging systems for modulation or control of the optical functionality. Even simple modulation and control of light of a given wavelength opens up applications in measurement and instrumentation through synchronous detection. Effective LCDs outside of visible wavelengths would enable modulation of signals for synchronous detection, reconfiguration of the optical system or wavelength tuning with lower power operation and the avoidance of mechanically moving parts. Some examples of these active research areas are given to illustrate the breadth of application of LCD technologies.

Of course, substrates and electrodes will need to be selected to be transmissive at the wavelength of operation, for example the use of fused silica by Sahoo *et al.* [15]. For some materials, the substrate itself may have both sufficient transparency and conductivity, such as the use of silicon substrates in the infrared.

8.4.2.3 Shorter wavelengths UV and deep UV(C)

One approach to exploiting liquid crystal molecular reorientation at wavelengths where the refractive indices are not themselves useful for a birefringence device is to include functional molecules or particles, as shown black shaded in Figure 8.11,

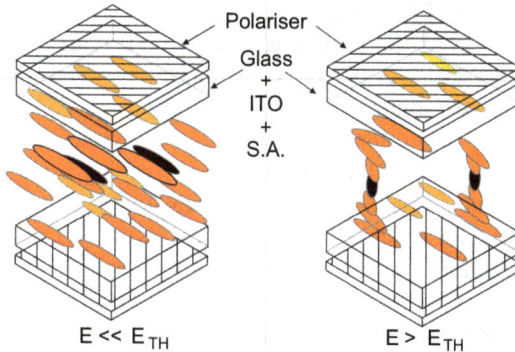

Figure 8.11 UV guest–host device configuration

that will reorient with the molecular motion in the liquid crystal phase. Polarisers may or may not be required depending on the application. This mechanism has been used in the past with dye molecules that possess anisotropic absorption. Reorientation of the dye molecules changes the absorption cross section of the device and has been used to produce a coloured device that does not require polarisers. For display devices in the visible region, the contrast achieved was not generally considered sufficient, and at high concentrations, the dye molecules may disrupt the alignment of the liquid crystal phase. However, for other applications outside of visible wavelengths, this offers a useful device-switching mechanism. Young *et al.* [16] have demonstrated the use of inorganic particle inclusions to modulate UV wavelengths down to 266 nm through this mechanism.

8.4.2.4 Longer wavelengths near-infrared (NIR) and far-infrared to terahertz frequencies

At longer wavelengths than the visible region, there are potential applications in thermal imaging and identification of materials through spectroscopic 'fingerprinting'. There is considerable interest and ongoing research in the use of liquid crystal materials and devices at these wavelengths. Many existing liquid crystal research materials are useable at NIR wavelengths. In fact, it has been broadly noted that liquid crystals can have useful refractive indices at wavelengths into the terahertz region of the spectrum. For example, E7, a liquid crystal research material from Merck commonly used at visible wavelengths, has been shown to have sufficient birefringence for useful LC devices at wavelengths around 1000 μm.

8.4.2.5 Near-infrared at wavelengths 1–3 μm

The NIR region of the spectrum is finding ever-increasing spectroscopic applications. Extraction of specific material data from NIR spectra is not so straightforward as for spectroscopy at longer wavelengths due to broad, overlapping spectra. Applications of optical devices include medical diagnostic and monitoring devices plus chemical quality control, including pharmaceuticals, foodstuffs and agricultural materials.

To illustrate some of the ongoing research at these wavelengths, a number of examples are cited from the literature. Xiaoxue [17] uses a cholesteric dual-frequency liquid crystal with bistability to switch between scattering and non-scattering states in the infrared. Sung *et al.* [18] have demonstrated a device working in the 1250–1650 nm range. This uses a hybrid construction with the changing refractive indices of the liquid crystal, providing a tuning mechanism for the transmission peaks of a diffracting structure. Similarly, Werner *et al.* [19] present the use of liquid crystal refractive indices to tune the refractive index of a metamaterial structure at wavelengths in the NIR region. The tuneable refractive index range actually includes positive and negative refractive index values.

Gutierrez-Cuevas [20] used a device construction similar to a guest host device. The guest material in this case is soluble gold nanorods. These couple to illumination in the NIR, and the energy induces a phase change in the liquid crystal that might be used for an optical switching effect. With such a device, there is a mechanism to directly optically convert NIR information to other wavelengths such as visible images or wavelengths optimised for a particular detector. The effect has been shown to be reversible on removal of the illumination.

One of the limitations of the use of liquid crystal materials in the infrared generally is the molecular resonances that give large absorption peaks, significantly reducing device transmission. Alternative materials with different structures are the subject of ongoing research [21], but with present structures, the bonds required for liquid crystal phases to exist inevitably have some absorption resonances in the 2–6 μm wavelength range [22].

8.4.2.6   Mid-infrared wavelengths 3–30 μm

In the Mid-infrared region, the thermal emissions of materials themselves become significant, and at wavelengths around 7–14 μm, detection of these emissions is the basis of thermal imaging systems. These images do not require artificial illumination so are effective for night vision, but can lack detail in extended areas of a scene at similar temperatures and with similar emissivities.

A number of types of LCD have been demonstrated in this wavelength range, including shutters and tuneable filters. There are inherent problems due to the absorption of the liquid crystal itself in the 3–5 μm region of the spectrum due to bond resonances. New liquid crystal materials are actively being researched to reduce absorption in this region [23].

8.4.2.7   Far IR wavelengths 30–1000 μm and terahertz

The terahertz wavelength region is something of a bridge between the millimetric waveguide technologies and the long-wave infrared photonic bandgap semiconductor emitter and detector devices. For inspection, sensing and screening, this region has some significant advantages. Many light materials such as clothing and packaging are transparent at these wavelengths, but the wavelengths are short enough to give reasonable spatial resolution, allowing identification of more electromagnetically dense objects. Systems at these wavelengths are directly applicable to security uses but are also likely to find application in process control and inspection.

For a birefringence device to have sufficient thickness for a $\lambda/4$ retardation $\Delta n.d$ at millimetric wavelengths, a device with a very thick liquid crystal layer is likely to be required. For devices with a liquid crystal layer of more than a few tens of micrometres, the alignment may begin to be disrupted by thermal fluctuations. Sahoo *et al.* [15] have demonstrated TN devices, but to operate at terahertz wavelengths, the thickness of the liquid crystal layers were 250 and 550 µm. Directly due to the material properties and thickness of the liquid crystal layers, the response times of some of these devices were of the order of 120 s. Although slow for a device to be used as part of a data acquisition cycle, this could be used for adjustment or reconfiguration of an optical path. Improvements could be obtained through the use of a dual-frequency liquid crystal to actively drive the liquid crystal orientation both planar and perpendicular to the field or to combine the birefringences of thinner devices.

Existing liquid crystal materials have useful birefringence from visible wavelengths through to terahertz wavelengths. The liquid crystal material available from Merck E7 has been the subject of a number of studies [24] of terahertz E7 properties and there are ongoing studies to improve liquid crystal properties for the terahertz wavelength region. Higher birefringence would allow for a thinner device, making alignment easier to maintain and to enable reduced switching times.

A currently active research area at a range of wavelengths is the use of matrices of resonators and similar structures to create materials with artificially controlled refractive indices. It is possible to create structures with negative refractive indices, e.g. Werner *et al.* [19]. At terahertz wavelengths, these structures can be fabricated using photolithography and ion beam techniques, although these can be expensive [25]. A promising form of construction for controlled devices is to combine these structures with a liquid crystal fluid that might either be electrically, optically or magnetically controlled [25].

8.4.3   Longer term future application areas

Photonic and optical systems offer non-contact methods of obtaining information through imaging and instrumentation and also physical interactions with subjects through the interaction of light beams with matter. In these respects, they are finding ever wider applications. As liquid crystal and closely related micro-fluidic devices and systems are developed for an ever-wider range of wavelengths, new application areas will open up. What follows are only a few key, speculative areas where photonics and optics are envisioned to develop and some examples of potential future systems.

8.4.3.1   Displays

Upon the advent of practical liquid crystal displays in the form of TN devices, there was at the time discussion and speculation of the concept of wall-mounted flat television and similar displays. However, it required decades of investment and innovation for them to become practical, affordable and of good image quality. This scopes the level and duration of development and investment required to take future possible applications from research concepts to good quality products.

There is always a drive to improve display functionality and quality. One area that is technologically demanding but receives attention because of the potential benefits is the display of depth information. Stereoscopic viewing technologies, taking images with an appropriate angular separation and displaying these images to left and right eyes work reasonably well for the majority of users. However, prolonged use can lead to eye strain and discomfort because of the disparity between the apparent distance due to the image angular separation and the distance at which the eyes are required to focus. Gaze tracking and adjustment of the focal distance with variable focus lens systems is a current research topic, though even then, other visual cues can cause disparities and discomfort. A volumetric display avoids such difficulties.

8.4.3.2   Volumetric displays

There are a number of applications where visualisation of volumetric information is extremely important such as surgical reconstructions and 3D design visualisation. A volumetric display (literally a volume in which 3D images are created) does not suffer from the artefacts of stereoscopic displays but does require some kind of physical material in the image volume. This material might emit light itself or scatter light from a light source outside of the display volume in order to form an individual illuminated point voxel (3D image basic element). For example, the display might take the form of an apparently transparent cube or cylinder. If the cylinder is directionally illuminated from above and/or below, then any voxel that is scattering will appear as a source of light to the viewer. However, using light scattering is likely to give some shadowing effects. Alternatively, a mechanism such as fluorescence might be used with non-visible wavelength illumination, and the voxels could employ fluorescent behaviour to appear as sources of visible light. There may still be some shadowing effects, but these may be reduced depending on the proportion of illumination absorption throughout the volume and the illumination pattern.

There are many problems to be solved in order to make a display of this type achieve good visual quality. One of the main optical problems is residual light scattering giving a background haze. This can be addressed through careful selection of material refractive indices, which will require material development and optimisation. There is also the question of the addressing mechanism for the voxels. There are existing technologies for transparent electrodes using ITO, but the electrodes would need to run throughout the volume of the display and this necessarily produces some absorption that would detract from the display appearance. Also, passive matrix addressing of such a large number of voxels (due to the extra dimension relative to flat pixel displays) would inevitably require a material with a very steep switching voltage transition or a bistable mechanism with a fast transition. The deployment of active-matrix technology within the display volume would obscure the displayed image itself.

A speculative optically based alternative mechanism is shown in Figure 8.12. Materials that respond to illumination at one wavelength to induce a conformation or phase change have been demonstrated. A similar mechanism might be used to

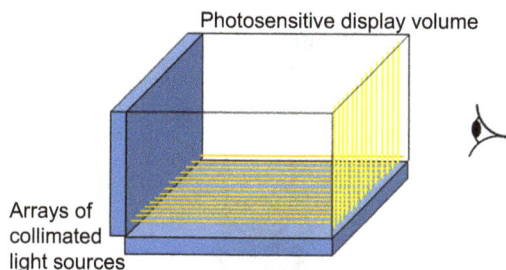

Figure 8.12   Volumetric display geometry

locally control fluorescence or light scattering in the display volume. Intensity-driven addressing would require confocal optics to achieve a distinct illumination point. Optical volumetric addressing is also likely to encounter problems similar to those of passive matrix addressing, with cross talk between voxels unless there is a clear threshold of response to the addressing intensity. Dual wavelength addressing might overcome this through a suitable combination of materials with a phase change that is only initiated when both illumination wavelengths are present. Only when and where both writing beams were present would there be a change of state. In this case, two sets of collimated beams could be used to initiate the phase change, with the intersections being swept through the display volume with associated intensity variations.

These solutions would require significant investment in materials and design, but the result would be a directly viewable volumetric display.

Some attempts at volumetric scattering displays using projection onto layered switchable scattering panels [26] and moving screens [27] have been demonstrated, but the number of screens required is large for high-depth resolution, and rapid movement of mechanical components is needed for projection onto screens.

8.4.3.3   Non-display applications

The non-display applications of photonic systems are many and varied, if not so apparent as their use as displays. Due to the versatility of LCDs, they can be applied in different ways, controlling illumination or selecting elements of the incoming light for instrumentation. The following sections outline some of the techniques and application areas under development that might become mainstream devices or applications in the coming decades. Some of these application areas may not see a eureka moment or step change but rather incremental improvements in resolution, signal-to-noise ratio and so forth, much as LCDs have incrementally progressed from the small visible light displays to much larger displays and now to other wavelengths.

8.4.3.4   Structured light

While structured light is not an application in its own right, it is a technology that finds application in a number of areas. It is an area where developments in LCD technologies are likely to find future applications. Structured light can take

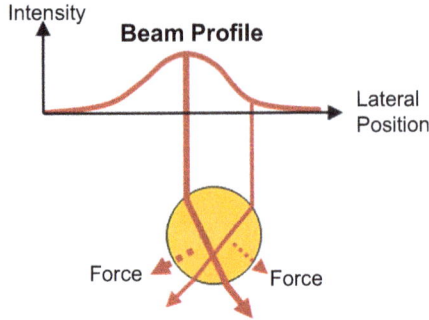

Figure 8.13   Dielectric particle beam coupling and trapping geometry

different forms, but the ability of LCDs to control light can be used in the generation and enhancement of the light patterns for these applications.

In its simplest form, structured light can be a spatial illumination pattern such as a grid, as part of image acquisition for 3D profiling. At a more detailed level, the light within a beam itself may have structure both in the intensity across the beam and its polarisation profile. If a laser beam is focussed on a beam waist at the diffraction limit, the intensity distribution created has an energy minimum for a dielectric particle at the centre of the focus. Figure 8.13 illustrates the restoring force on an off-axis particle. This is the essence of optical tweezers. The particles may be material samples or biological materials such as individual cells [28]. Trapping of multiple particles has been demonstrated, as has the transfer of angular momentum to induce rotation of the particles. Control of the position and orientation facilitates applications such as spectroscopy on individual particles or cells or the sorting of cells in biological processes.

The ability to physically handle and manipulate small dielectric particles such as cells, fluid droplets or other tiny objects (10nm to 10,000nm) is a valuable tool in microscopy and physical characterisation, but currently, these are complex laboratory-scale instruments and prohibitively expensive.

Polarisation control is still a key development. Asymmetric particles either physically or through birefringence will rotate in circularly polarised light. Optical vortices may be generated through holographic waveplates. Control of these optical functions would allow more precise control of the particles' position, orientation and rate of rotation. LCDs may be used directly to control and manipulate light polarisation or to modulate the effect of metasurfaces such as lenses or holographic beam shaping.

In the longer term, for these instruments to become more compact and affordable, the use of fibres and metasurfaces for light profiling is a subject of current research [28]. Liquid crystal or micro-fluidic devices would permit reconfiguration of the light pathways through the instrument.

Scaling-up optical trapping from visible and infrared wavelengths to the terahertz region would potentially give a controlled volume of millimetric rather than micron dimensions. Especially in a freefall or micro-gravity environment this could be used to manipulate and manoeuvre small objects.

8.4.3.5 Optical information processing

Optical information processing is signal processing through optical means. Optical information processing systems do offer whole image operations in parallel and straightforward transforms to and from Fourier space where filtering of the spatial image frequencies may be carried out. They also offer easy interconnectivity, as the optical signals can pass through the same space without interference. With switchable optical components, these optical functions may be reconfigured to be carried out at different wavelengths or to filter different spatial frequencies, etc. Spatial light modulators (SLMs), which are in effect miniature liquid crystal displays, are commonly used as input devices to optical information processing systems.

One of the features embodied in electronic image processing is the ability to apply non-linearity to the data. Non-linear relationships are not easily applied in an all-optical photonic system. There has been significant investment in achieving non-linearity in electro-optic crystals. However, inducing sufficient charge movement and re-distribution has been found to require significant optical pump powers and applied electric fields to induce useable changes in optical properties such as the hyperpolarisability (chi).

Due to the sensitivity of LC phases to molecular morphology, there are mechanisms for optically induced liquid crystal phase changes. These and the response of liquid crystals to relatively low externally applied fields provide potentially useful switching mechanisms including hysteresis and thresholding. Application of some of the optically induced liquid crystal switching mechanisms to optically controlled devices would provide optical feedback and facilitate non-linear operations.

Optical neural network computing architectures were demonstrated using optically addressed SLMs (OASLMs) in the 1980s and 1990s. In the longer term some of the optically driven optical function interactions open up the possibility of a new generation of optically addressed spatial light modulators of simplified construction and not necessarily requiring electrical bias fields. Using optically controlled optical devices would facilitate an all-optical system, eliminating conversions between the optical and electronic regimes.

This work continues to develop apace and it is likely that optical information processing will find applications in areas where there are specific advantages to this type of processing. This is especially true of computational imaging, where the final 'image' is the product of different sources of data that have to be merged.

In a hybrid computational imaging system, the optical part may pre-process the image to extract essential features that may be used to inform an application. This could be similar to the way biological visual systems extract triggers from images for horizontals, verticals, etc. Optical processing might 'pre-process' images to enhance or extract features prior to electronic computational image analysis. This might be wavelength selection, spatial transforms or extraction of features such as edges in advance of assimilation into the electronic computational domain. These are likely to be incorporated in embedded applications such as vision systems for autonomous vehicles and robotics or instrumentation systems for process and environmental monitoring, effectively enhancing digital eyes.

At present autonomous vehicles such as the Perseverance rover operate on Mars at limited speeds with elementary stereoscopic imaging. Enhancements to the imaging processing bandwidth through optical pre-processing could improve obstacle identification allowing more reliable operation at higher vehicle speeds.

8.4.3.6   Sensing and instrumentation

Optical and photonic sensing instrumentation provides contact-free and remote methods of obtaining a range of physical measurements. Spectral or multi-wavelength measurements are already widely exploited in environmental and atmospheric monitoring, wind velocity measurements and spectral pollution monitoring. The reconfiguration and control of optical systems offers selection from optical functions, polarisation or wavelength selection or weight and volume savings compared to bulky mechanical reconfiguration mechanisms.

The birefringence of liquid crystals, often seen as a disadvantage in that some of the light goes to waste, can be used to advantage in selecting the pathway through an optical system or modifying optical functions with birefringent optical components [14].

Photonic and optical instrumentation already finds extensive application in environmental and process monitoring. These applications will benefit from progressive improvements in system performance. One particular area receiving a lot of attention is the development of measurements for healthcare.

8.4.3.7   Healthcare

Ongoing monitoring and screening for healthcare offers big advantages in health with earlier detection and consequently better treatment outcomes. Ideally, these measurements should be carried out in a non-invasive manner. Some of the earliest medical optical devices were of course eyeglasses, and optical retinal scans are now a routine high street activity and tissue oxygen saturation levels are easily measured optically as a ratio of transmission through tissues at different wavelengths. Adaptive optics approaches to ophthalmic imaging allow significantly enhanced resolution microscopy. Though something of a laboratory instrument at present, ultimately this might be rolled out to the high street for routine screening.

The ultimate goal of course would be an external non-contact device, handheld or possibly wearable, that would be capable of scanning for a wide range of biological functions. With the reduction in the size of complex optical systems, more detailed optical measurements become wearable or possibly even implantable.

Many photonic medical screening processes at present involve laboratory-scale tests on samples taken. Even the present wearable glucose monitors depend on an embedded sensor with only a finite lifetime. As optical systems are refined, they will open up the possibility of real-time in vivo monitoring rather than in vitro testing. Working at wavelengths further from the visible, going deeper into the longer wavelengths, and also to shorter wavelengths, will open up new applications. Longer wavelengths penetrate deeper into tissues, thus enabling internal, in vivo spectroscopy. Shorter wavelengths probe fluorescence mechanisms. Multi-spectral imaging across different wavelength ranges and data synthesis has the

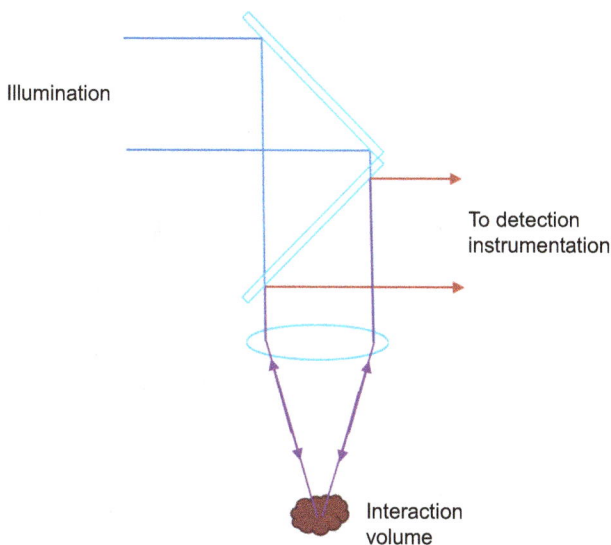

Figure 8.14   Confocal backscatter measurement geometry

potential for cross-checking and cross-correlating results. Potentially these methods can be used for the detection of the presence of pathogens or the metabolic products of specific conditions in blood and tissues. Optical blood sugar monitoring is common but still effectively requires sampling, either with a finger prick test or an embedded worn sensor.

The use of LC technologies permits frequency selection without the use of voluminous gratings. If more precise depth resolution is required, confocal geometries such as in Figure 8.14 may be used to target a specific sample volume and, through the use of variable focus lenses, this sample volume might be swept or selected.

For some of these healthcare technologies to become practical, there need to be parallel developments in both biochemistry and optical technologies.

The earliest optical healthcare application was vision correction with eyeglasses, and recently, liquid crystal contact lenses are being developed that provide variable accommodation correction [29].

8.5   Summing up

In this chapter, it has only been possible to present an overview of the potential of liquid crystal photonic and optical technologies. There are some specific liquid crystal phases and switching effects that have not been treated in detail: flexoelectric coupling, for example [30,31].

There are also photonic application areas that would require a chapter in their own right. For example, quantum computing and communications using entangled photons is an area beyond the scope of this chapter but one that will benefit from the optical control possible with LCD technologies. Overall, the future applications of LCD technologies are likely to be as diverse as the myriads of liquid crystal phases that have been shown to exist.

References

[1]   Collings P.J. and Hird M. *Introduction to Liquid Crystals: Chemistry and Physics*. Milton Park: Taylor and Francis; 1997.

[2]   Boller A., Scherrer H., Schadt M., and Wild P. 'Low electroptic threshold in new liquid crystals'. *Proceedings of the IEEE*. 1972;60(8): 1002–1003.

[3]   Gooch C. and Tarry H., 'Optical characteristics of twisted nematic liquid crystal films'. *Electronics Letters*. 1974;10(2): 188.

[4]   Raynes E.P. and Waters C.M. 'Supertwisted nematic liquid crystal displays (review)'. *Displays*. 1987;8(2): 59–63.

[5]   Gleeson H.F.G. *Thermography Using Liquid Crystals. Handbook of Liquid Crystals Set*. New York: Wiley; 2008.

[6]   Lagerwall S.T. 'Ferroelectric liquid crystal displays with greyscale'. *Liquid Crystals Today*. 1996;6(2): 5–7.

[7]   Lee C. S. 'An electrically switchable visible to infra-red dual frequency cholesteric liquid crystal light shutter'. *Journal of Materials Chemistry C* 2018;6 (15): 4243–4249.

[8]   Lester G., and Strudwick A. 'Instrumentation wavelength range selection and reconfiguration using electrically switchable diffraction gratings'. *Proceedings of SPIE*. 2023;12442: 124420F.

[9]   Lester G. 'Optoelectronic devices for reconfigurable imaging and optical systems' in Pandalai S. (ed.). *Recent Research Developments in Electronics* Vol. 1. Trivandrum: Transworld Research Network; 2002. pp. 165–175.

[10] Mc Owen P.W., Gordon M.S., and Hossak, W.J. 'A switchable liquid crystal Gabor lens'. *Optics Communications*. 1993;103(5–6): 189–193.

[11] Hudson T., and Gregory D.A. 'Optically-addressed spatial light modulators'. *Optics & Laser Technology*. 1991;23(5): 297–302.

[12] Yue Y., Norikane Y., Azumi R., and Koyama E. 'Light-induced mechanical response in crosslinked liquid-crystalline polymers with photoswitchable glass transition temperatures'. *Nature Communications*. 2018;9: 3234.

[13] Kaan A., Baghsiahi H., Surman P. *et al.* 'Dynamic exit pupil trackers for autostereoscopic displays'. *Optics Express*. 2013;21(12): 14331–14341.

[14] Love G.D., Hoffman D.M., Hands P.J.W., Gao J., Kirby A.K., and Banks M.S. 'High-speed switchable lens enables the development of a volumetric stereoscopic display'. *Optics Express*. 2009;17(18): 15716–15725.

[15] Sahoo A.K., Yang C.-S., Yen C.-L. *et al.* 'Twisted nematic liquid-crystal-based terahertz phase shifter using pristine PEDOT:PSS transparent conducting electrodes'. *Applied Science*. 2019;9(761): app9040761.

[16] Young H.L. '2D material inorganic liquid crystals for tunable deep UV light modulation'. *National Science Review*. 2022;9(12): nwac252.

[17] Xiaoxue D. 'Electrically switchable bistable dual frequency liquid crystal light shutter with hyper-reflection in near infrared', *Liquid Crystals*. 2019;46(11): 1727–1733.

[18] Sung G.F., Chiu S.-Y., Chang Y.-C., Liou Y.-C., Yeh C.-P. and Lee W. 'Electrically tunable defect-mode wavelengths in a liquid-crystal-in-cavity hybrid structure in the near-infrared range'. *Materials*. 2023;16(3229): ma16083229.

[19] Werner D.H., Kwon D.-H. and Khoo I.-C. 'Liquid crystal clad near-infrared metamaterials with tunable negative-zero-positive refractive indices'. *Optics Express*. 2007;15(6): 3342–3347.

[20] Gutierrez-Cuevas, K.G., Wang L., Xue C., *et al.* 'Near infrared light-driven liquid crystal phase transition enabled by hydrophobic mesogen grafted plasmonic gold anorods'. *Chemical Communications*. 2015;51(48): 9845–9848.

[21] Kula P., Bennis N., Harmata P., *et al.* 'Perdeuterated liquid crystals for near infrared applications', *Optical Materials*. 2016;60: 209–213.

[22] Harmata P. and Herman J. 'New-generation liquid crystal materials for application in infrared region'. *Materials*. 2021;14(10): 2616–2631.

[23] Chen Y. 'Low absorption liquid crystals for mid-wave infrared applications'. *Optics Express*. 2011;19(11): 10843.

[24] Yang C.-S., Lin C.-J., Pan R.-P., *et al.* 'The complex refractive indices of the liquid crystal mixture E7 in the terrahertz frequency range'. *Journal of the Optical Society of America B*. 2010;27(9):1866–1873.

[25] Wang L., Wang Y., Zong G., Hu W., and Lu Y. 'Liquid crystal based tunable terahertz metadevices', *Journal of Materiomics*, 2025;11(1): 100888.

[26] Sullivan A. 'DepthCube solid-state 3D volumetric display'. *Proceedings of SPIE*. 2004;5291: 5291: 279–284.

[27] Favalora G. E., Napoli J., Hall D. M., *et al.* '100 million-voxel volumetric display'. *Proceedings of SPIE*. 2002;4712: 300–312.

[28] Volpe G., Marago O.M., Rubinsztein-Dunlop H., *et al.* 'Roadmap for optical tweezers'. *Journal of Physics: Photonics*. 2023;5(2): 1–135.

[29] Bailey J., Kaur S., Morgan P.B., Gleeson H.F., Clamp J.H. and Jones J.C. 'Design considerations for liquid crystal contact lenses'. *Journal of Physics D: Applied Physics*. 2017;50(48): 485401.

[30] Fells J.A., Welch C., Wing C., *et al.* 'Dynamic response of large tilt-angle flexoelectro-optic liquid crystal modulators', *Optics Express*. 2019;27(11): 15184–15195.

[31] Coles H., Musgrave B., Coles M., and Willmott J., 'The effect of the molecular structure on flexoelectric coupling in the chiral nematic phase'. *Journal of Materials Chemistry*. 2001;11(11): 2709–2716.

Chapter 9

Novel and advanced particle accelerators

Graeme Burt[1], Laura Corner[2] and Rebecca Seviour[3]

Particle accelerators are best known for their uses in high-energy physics in places like CERN; however, there are tens of thousands of particle accelerators all around us. Small-scale accelerators are used in applications like treating cancer, scanning cargo at ports, cross-linking polymers, colouring gemstones and curing paints and composites, whereas larger-scale accelerators generate intense X-ray or neutron beams to enable pharmaceutical, chemical, archaeological, biological and solid-state physics research or analysis [1]. In the past decade, there has been renewed interest in new accelerator technologies that can reduce the size or increase the performance of particle accelerators, opening up new avenues of research by making large facilities to fit in a university laboratory, performing higher resolution scans or improving cancer treatment. In this chapter, we will explore three technologies capable of delivering a step change in accelerator design: new super-conducting materials and structures, higher frequency particle accelerators, and plasma-based particle accelerators.

9.1 Introduction to accelerators

The first particle accelerator was created by Cockcroft and Walton in 1932 [2] in order to split a lithium atom with protons, for which they received the Nobel Prize. Since then, physics experiments have required ever higher energies, ranging from 400 keV in the first Cockcroft–Walton accelerator to 13.6 TeV in the current Large Hadron Collider (LHC) at CERN. Future experiments will require energy greater than 100 TeV. However, the accelerating gradient, or the energy gained by the particles per meter, has only increased from 1 to 100 MV/m since the first linear accelerator in 1934. This has meant that accelerators for high-energy physics have had to increase dramatically in size from a scale of a few metres up to tens of kilometres. Future accelerators will need to operate at higher accelerating gradients to increase the collision energy without increasing the size of the accelerator.

[1]Engineering Department, Lancaster University, UK
[2]Engineering Department, University of Liverpool, UK
[3]Engineering Department, University of Huddersfield, UK

There are many other applications for particle accelerators. One important use is in radiotherapy for cancer treatment. There is significant interest in using particles for radiotherapy rather than X-rays as the dose can be better targeted on the tumour, sparing health tissue surrounding it. However, using particles such as electrons [3], protons or ions [4] requires energies more than 100 MeV in order to achieve sufficient penetration, which increases the size of the accelerator. Accelerators are also used for making neutrons or X-rays for the study of medicines, materials, viruses and archaeology. The use of synchrotron light sources was critical to the development of coronavirus vaccines, for example [5]. The size of these machines requires them to be at one or two national centres per country. It is therefore clear that there are immediate advantages of increasing the accelerating gradient of particle accelerators so that they can be reduced in size and hence placed in hospitals and laboratories all around the world.

New technologies are currently in development that can significantly exceed the gradient of existing accelerators. The main limit in gradient is a phenomenon known as vacuum breakdown, where a plasma is formed inside the accelerating structures, so these technologies seek to delay or avoid such events [6]. A major increase in gradient in the past 20 years has come from increases in the operating frequency of particle accelerators from 3 to 12 GHz as the electric field at which breakdown occurs scales upwards with frequency. However, increasing the frequency also reduces the size of the structures, making them difficult to manufacture and reducing the amount of beam current that can be accelerated. Further increasing the frequency up to 300 GHz would allow a significantly higher gradient, as well as reducing the transverse size of the accelerator, if these issues can be overcome. Another way to generate higher gradients is to accelerate inside a plasma such that breakdown is no longer an issue. Here, the accelerating fields are driven by either a laser or another particle beam. However, plasmas are inherently unstable. Making miniature accelerators could open a whole range of applications of accelerators that are not currently possible such as portable X-ray machines for security or bomb disposal, removing toxic chemicals from water on a mass scale, reducing chemical uses in industries such as tanning leather by using electrons instead or robotic radiotherapy systems with full range of movement using electrons or protons.

In addition to energy and gradient, there are many other important requirements for particle accelerators to be useful. The beam quality has to be sufficient for the application. In many cases, this means that all the particles must have the same energy within a fraction of a percent, and they need to be focused on a small spot. The beam should be repeatable and reliable. For example, in radiotherapy, an error in energy or focus would result in a large radiation dose being deposited on healthy tissue, leading to cell damage or the risk of secondary cancers. The beam current is also important to minimise the accelerator time required, no matter the application. This results in more collisions for high-energy physics research, shorter treatment times in radiotherapy and shorter imaging times in pharmaceutical studies. Indeed, several high-gradient plasma accelerators are already able to deliver GeV beams in a few centimetres but do not yet have sufficient repeatability for most practical applications.

Finally, we need to care about the sustainability of particle accelerators. The LHC uses 600 GWh of power every year, and future accelerators could require significantly more if new technologies are not implemented to minimise the power usage. The LHC would use significantly more power if it were not for the use of superconducting cavities [7]. Superconductors, when cooled below a certain temperature, allow currents to flow with little or no resistance and, hence, generate significantly less heat than normal conducting structures. This, in turn, means less energy loss and higher efficiencies. However, superconducting cavities are limited to a gradient of 10–30 MV/m due to the breakdown of superconductivity at high fields, known as quenching and heating from field-emitted electrons hitting the surface [8]. To make viable accelerators in the future, the gradient of super-conducting cavities must be increased significantly.

9.2 Superconductivity and particle accelerators

9.2.1 Background

In metals with high conductivity, such as copper, the resistivity decreases with temperature until reaching a constant ($\sim 10^{-9}$ $\Omega \cdot$ m) at around 20 K (Figure 9.1). Although, in 1911, Onnes found that by cooling mercury, considered a bad conductor, below 4.16 K, the resistivity suddenly vanished, becoming a superconductor at this critical temperature [9]. In addition to this critical temperature (T_c) where the resistance of the material disappears there is also a critical magnetic field (H_c). If $T < T_c$, then, in the presence of an applied magnetic field (H), the resistance of the material disappears when $H \leq H_c$ and is finite for $H > H_c$. Furthermore, for $H \leq H_c$, the magnetic field is expelled from the superconductor.

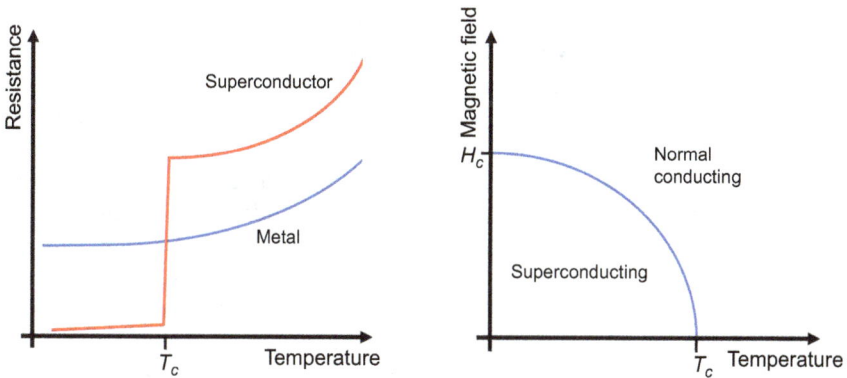

Figure 9.1 *(Left) Graph of resistance vs. temperature for superconductors and normal metals. Right panel showing the point of onset of superconductivity of temperature vs magnetic field for a type-I superconductor.*

This exclusion of the magnetic field from a superconductor is termed the Meissner effect. It should be noted that the Meissner effect is a property of the superconductor and not due to zero resistance. This is a key difference between a superconductor and a perfect electrical conductor (PEC), as PEC materials do not expel an applied magnetic field.

One explanation for the phenomenon that gives rise to superconductivity was proposed by Cooper [10]. Key to this theory is the realisation that vibrations in the atomic lattice of certain materials (phonons) can mediate an attractive interaction between two electrons traveling through the lattice of the material [11]. As an electron moves between the positively charged nuclei of the material's lattice, it exerts an attractive force on the surrounding (large, relatively stationary) atomic nuclei, creating small spatial distortions in the distribution of the positively charged nuclei. These spatial variations cause small localised regions within the lattice to gain a positive charge, which attracts an additional electron to the region of the spatial distortion. This gives rise to a coherent motion between the two electrons, as one electron moves, the distortion in the lattice moves with the electron, pulling the other electron with it, giving rise to the coherent collective motion of the two electrons, called a Cooper pair. The length of this coherence (the mean separation), or correlation length, between electrons in a Cooper pair is material dependent, ranging between 100 and 1000 nm. Much larger than the sub-nm separation between atoms, meaning Cooper pairs overlap with millions of other Cooper pairs in a superconductor. This overlapping inexorably results in a highly collective condensate. In this condensate, the breaking of one pair will change the energy of the entire condensate, i.e. the energy required to break a single Cooper pair is related to the energy required to break all Cooper pairs. This energy barrier means collisions by the Cooper pairs with lattice defects or impurities are insufficient to affect the condensate as a whole (or breaking coherent pairs), meaning that the current flow carried by Cooper pair excitation's in a superconductor will not experience resistance. This initial model was expanded by Bardeen, Cooper and Schrieffer into BCS theory [12,13], which predicts this phase transition, the Meissner effect and the penetration depth [12,13].

Further research yielded the discovery of multiple materials that exhibit superconductivity, including chemical elements (mercury, lead, niobium), alloys (e.g. niobium–titanium, germanium–niobium, and niobium nitride), ceramics (yttrium barium copper oxide and magnesium diboride), and superconducting pnictides (fluorine-doped LaOFeAs). As superconductivity is a thermodynamic phase, certain physical properties of superconductors are largely independent of specific material structure, such as the Meissner effect, the quantisation of the magnetic flux or permanent currents all of which are superconducting-material independent. Other physical properties, such as the critical temperature, the value of the superconducting energy gap, the critical magnetic field, and the critical current density at which superconductivity is destroyed, vary from material to material. The variation in physical properties of superconducting materials can be broken down into two types of superconductors Type I and Type II. Type I and Type II superconductors primarily differ in their response to an external magnetic

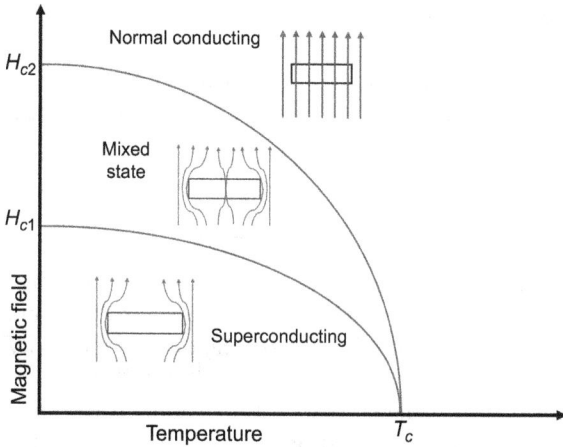

Figure 9.2 On set of superconductivity of temperature vs. magnetic field for a type-II superconductor, showing the magnetic field lines for the three configurations

field and their critical magnetic field values. Type I superconductors, typically composed of pure elements like lead or mercury, exhibit a complete transition to the superconducting state below a certain critical temperature (T_c) and a critical magnetic field (H_c). As discussed previously, when the applied magnetic field exceeds H_c, Type I materials revert to a normal, non-superconducting state. In contrast, Type II superconductors, often made of alloys or complex compounds such as niobium-titanium or yttrium barium copper oxide, transition to a super-conducting state below Tc, but exhibit two critical magnetic fields, H_{c1} and H_{c2}. Between these fields, they enter a mixed state where magnetic vortices penetrate the material, allowing partial superconductivity to persist even under higher magnetic fields (Figure 9.2). This mixed state enables Type II superconductors to maintain superconductivity in much stronger magnetic fields compared to Type I superconductors, making them more suitable for some practical applications where strong magnetic fields are present, such as the RF cavities of particle accelerators.

9.2.2 Resistance

As discussed, the current carried by Cooper pairs in a superconductor flows without resistance, as the collision of Cooper pairs with lattice defects or impurities is insufficient to break the coherent pairs. However, in the case of Type-II super-conductors we can have a mix of electrons in Cooper pairs and unpaired electrons. In the DC case only the lossless Cooper paired electrons contribute to the current, as the Cooper pairs expel EM fields in the superconductor, essentially screening normal conducting electrons from the applied electric field and hence they do not contribute to the current. In the presence of an externally applied RF field, the situation is more complex. In response to the RF field, the current carried by the Cooper pairs oscillates, and as the Cooper pairs posses momentum, they no longer completely screen the external EM fields. Allowing the non-paired, normal conducting, electrons to respond the applied EM field resulting in power dissipation. Although as the temperature of the superconductor decreases, the number of unpaired electrons declines exponentially, resulting in an exponential decrease in surface resistance.

9.2.3   Superconducting materials and particle accelerators

In particle accelerators superconducting materials are utilised in two key technologies. The first is the RF resonant cavity, a resonant structure designed to store EM energy at a specific frequency and EM field configuration, that oscillates an electric field parallel to a charged particle beam traveling through the cavity. By design the charged particle bunch is timed to pass through the RF cavity at an appropriate phase of the electric field such that energy is transferred from the EM field into the charged particles, accelerating the particle bunch. The second key application of superconducting materials in particle accelerators is magnets, to create high magnetic fields to steer and focus the accelerated charged particle beam in the accelerator.

9.2.4   Superconducting RF

RF power requirements are a critical factor in the design and running particle accelerators. As the cost of procuring and running RF sources increases dramatically with increasing power requirements. For normal conducting cavities, the power loss in the cavity walls can easily equal or exceed the beam power consumption. Whereas for superconducting cavities, wall loss is negligible, meaning nearly all RF power goes to the beam, allowing for a far higher efficiency compared to an NC accelerator.

The objective of an RF cavity is to transfer energy from the EM field stored in the cavity to the particle beam, achieved by the accelerating voltage ($V_a cc$) established by the EM field parallel to the particle beam. In any accelerator, it is important to optimise the efficiency of this energy transfer, a key design parameter used to asses this efficiency is the shunt impedance ($R_{sh} = V_{acc}^2/(2P_c)$), that relates the accelerating voltage to the loss of EM power, via induced currents, in the surface of the cavity walls (P_c). To maximise R_{sh} we need to minimise the surface resistance of the cavity. The negligible surface resistance of SC cavities leads to a difference in the R_{sh} between superconducting and NC cavities of order 10^5, making SC cavities a very attractive proposition for particle accelerator design.

Although the low surface resistance of superconducting materials naturally leads to very high R_{sh}, there are still a number of issues that have to be considered when designing superconducting cavities. Such as thermal loading, even though P_c is low the heat generated needs to be dissipated/kept-small to prevent quenching of the superconductor. One of the routes this heat is dissipated is through the liquid He coolant, introducing a further loss that impacts the efficiency of the RF system if not managed. Also, unlike normal conducting cavities, superconducting cavities have to be designed to minimise the EM fields on the cavity surface, i.e. to prevent the magnetic field on surface exceeding H_{sh} of the superconducting material. Hence, superconducting cavities are designed with elliptical surfaces and large beam pipe irises.

To appreciate the advantages of SCRF cavities, consider a particle accelerator with normal-conducting cavities and a high duty factor, or continuous operation. The oscillating EM field inside the cavity induces current to flow in the walls of the cavity, which results in losses. These electrical losses can be high enough that the material of the cavity melts (even in water-cooled high-conductivity Cu cavities). Whereas the negligible electrical loss of superconductors means SCRF cavities can operate at higher power with high duty-factor (or continuous operation) compared to normal conducting cavities. Although this does not mean SCRF cavities can be operated without consideration of loss, even small electrical losses can cause localised heating, that if not managed can cause thermal breakdown (quench) of the superconductor. Meaning care must be taken when designing the cryogenic system that supports the SCRF cavity to ensure thermal loading is managed.

Normal conducting cavities require geometries designed with small beam apertures to concentrate the accelerating electric field, to compensate for power losses in the cavity walls. This small aperture can result in particles passing through the aperture to create wakefields, a parasitic effect that can perturb subsequent particle bunches as they pass through the cavity. The low electrical loss in an SCRF cavities allows the geometry of the cavity to have a much large beampipe apertures compared to conventional warm technologies (Figure 9.3), and still maintain a high accelerating field along the beam axis, giving rise to systems with low beam-impedance and loss.

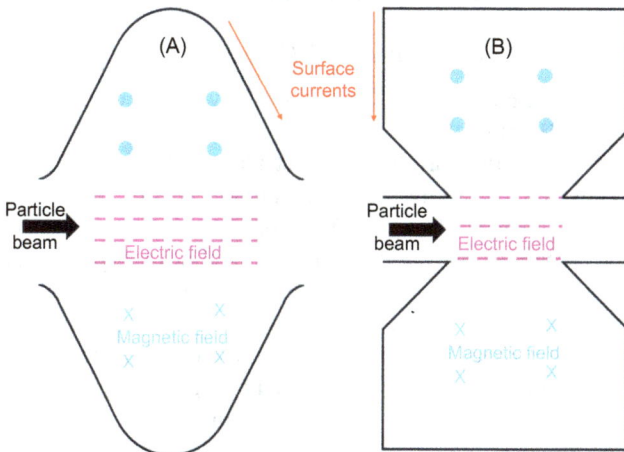

Figure 9.3   Schematic RF cavity cross section of (A) a typical SC RF cavity, (B) a typical NC, reentrant RF cavity. Showing the electric (horizontal) and magnetic (through the page) fields, surface currents and the direction of the particle beam.

A key factor in choosing the superconducting material to form an RF cavity from is the limits the material has to an applied magnetic field. When a static magnetic field is incident on a superconductor with a field strength greater than the critical field H_c, then the magnetic field starts to penetrate the superconducting material, creating normal conducting islands. In the case of an RF magnetic field current convention holds that the creation of normal conducting islands in the superconductor are formed on time scales greater than the RF period; the field at which this occurs is referred to as the critical superheating field H_{sh}, which is below 1 T for all known superconductors, but much larger than H_c (H_{c1}) of any known superconductors [14]. In addition, we should note that flux pinning is coupled with hysteretic losses, meaning a 'soft' superconductor is a good choice for a super-conducting RF cavity. These factors make Nb a good candidate to form super-conducting RF cavities [15], with a T_c of 9.2 K and a H_{sh} of 240 mT.

Figure 9.3 presents the schematic cross-section for both a typical super-conducting RF cavity and a typical normal conducting reentrant RF cavity. In the superconducting RF cavity design, note the large iris opening and a rounded shape to minimise the peak EM fields on the surface and to ensure the trajectories of any 'stray' electrons are prevented from achieving a resonant parasitic state. Whereas the normal conducting RF cavity is designed to optimised a high shunt impedance, with 'reentrant' nose cones at the points of entry of the particle beam. This reen-trant design reduces the volume that the energy of the EM field can be stored over, increasing the energy density in particular around the nose cones, leading to an increased electric field at the point required to accelerate a charged particle beam.

In summary, superconducting RF cavities offer three main advantages over warm RF cavity technologies. High-duty-cycle or continuous operation, low beam impedance, and high transfer of power to beam.

9.2.5   *Superconducting thin film RF cavities*

The EM fields in SCRF cavities only penetrate a few nanometers into the surface of the cavity (about 40 nm in Nb), i.e. the EM field only 'sees' the first micron of the cavity surface. Paving the way for low-cost alternatives to bulk Nb cavities, by depositing a superconducting film (typically around 2 μmT [16]) on the surface of a castable RF cavity made of Cu or Al. This approach dramatically changes the cost of SCRF based accelerators [16], and offers several advantages compared to bulk Nb cavities. The BCS surface resistance for the SC thin-films is lower than bulk materials, due to the normal state electrical resistivity being closer to optimum. The Cu substrate has higher thermal conductivity which results in increased temperature stability for thin-film SCRF cavities compared to bulk cavities. The micron thick-ness of the SCRF film dramatically reduces the amount of material needed compared to a bulk SCRF cavity resulting in large cost savings, especially for low-frequency cavities.

The performance of thin-film SCRF cavities depends on the uniformity of the thin-film morphology and density. Unfortunately RF cavities generally have com-plexed, curved, geometries, meaning the creation of viable thin-film SCRF cavities is non-trivial. The technology of choice for forming SCRF thin-film is magnetron sputtering (DCMS), where a cylindrical Nb cathode is placed inside the castable RF cavity [16]. Although the anisotropic fluxes DCMS creates result in preferential line-of-sight deposition creating geometrical shadows, meaning thin-films of *Nb* with uniform thickness and morphology is challenging [16].

A possible candidate deposition technology that over comes these limitations is high-power impulse magnetron sputtering (HiPIMS) [16]. Where power is applied to the sputtering cathode, in short unipolar pulses, with duty cycles shorter than 10%, allowing for power densities on the substrate surface to exceed kW cm^{-2}, creating plasma densities orders of magnitude higher than can be achieved by DCMS. The deposition of the ionised fluxes can be controlled by applying an electrostatic bias potential to the substrate such that ions can be deflected on to surfaces that do not face the magnetron source [16].

Although thin-film SCRF cavities exhibiting very high Q at low field they are not without issues, such as steep RF losses at medium to high gradients [16] extensive research programmes are underway across the world to understand and addresses these issues. Where successful development of thin-film SCRF cavities will have a significant cost reduction in cavities and cryomodule cost [16].

9.2.6 Alternative materials

Conventional bulk 1.3 GHz cavities reliably achieve quality factors of around 4×10^{10} and accelerating gradients of over 35 MV/m and 2 K, but improving cavity performance further is becoming increasingly difficult and exponentially expensive as we approach the intrinsic limits of Nb. In addition, conventional Nb SCRF cavities require sub-5K temperatures to operate requiring expensive cryogenic infrastructure, limiting access to SCRF technology. To address this high infrastructure outlay and running costs research is underway to identify alternative superconducting materials capable of operating at temperatures above 4 K, such as Nb_3Sn, considered by many as the SCRF material of the future [17] for particle accelerators, a material capable of operating at 18 K, removing the need for expensive and power hungry cryogenic infrastructure. Alternatively Nb_3Sn cavities operating at 4.5 K have the potential to achieve quality factors an order of magnitude higher than conventional Nb cavities (due to the higher T_c), and offers a theoretical pathway to cavities with 100 MV/m accelerating gradients [18].

The current approach to fabricating Nb_3Sn is to diffuse Sn into the surface of NbO, creating a Nb_3Sn thin film [17]. This diffusion of Sn requires high-temperature processing, which makes the Nb_3Sn film brittle and can result in high strain in the Nb_3Sn film that can destroy the superconducting state due to the strain sensitivity of the critical current [19]. Research is currently underway to fabricate Nb_3Sn thin film SCRF cavities by deposition [19] to get around these issues. Although there are many challenges to overcome before this material can be used in particle accelerators, particularly around the growth of the thin film, and developing our understanding of *Sn* adsorption and diffusion behavior on NbO [17]. There are of course a plethora of superconducting candidate materials of which Nb$_3$Sn is but one, others such as YBaCuO, BSCCO and MgB$_2$ have been considered, but issues with RF losses and the brittle composition has so far made such material unsuitable for application as RF cavities. Materials such as NbTiN, NbN, V$_3$Si have also been considered and are still the focus of research. Although future developments may very well incorporate these or other materials into future technologies.

9.2.7   Superconducting magnets

Particle accelerators use several types of magnet technologies (dipoles, quadru-poles, sextupoles) to guide and focus particle beams. Dipole magnets create a uniform magnetic field, bending the particle beams trajectory along a circular or curved path and maintaining the beam's orbit, either to store particle in a storage ring or to pass them repeatedly through an acceleration section of circular particle accelerator. Quadrupole magnets produce a gradient magnetic field with opposite polarity on either side to focus the particle beam in one plane whilst defocusing it in the perpendicular plane, keeping the beam tightly packed and minimising divergence. Sextupole magnets generate a more complex magnetic field that corrects higher-order aberrations in the beam's path, improving focus and stability by addressing non-linear distortions and chromatic aberrations.

Superconducting magnets are powerful electromagnets formed from coils of superconducting wire that when cooled below their critical temperature can carry very high currents without resistance. Generating exceptionally strong and stable magnetic fields, exceeding 45 T, far higher than those produced by conventional copper or iron-core electromagnets. The efficiency of these magnets stems from the zero electrical resistance of superconductors, eliminating the losses associated with the high currents, and allows continuous operation with minimal power. Despite the need for cooling to cryogenic temperatures the performance benefits and energy efficiency of superconducting magnets make them indispensable for large accelerators such as the LHC. This efficiency allows particle accelerators to maintain high magnetic fields continuously without prohibitive operational costs. Additionally, the compact size of superconducting magnets, due to their high current-carrying capacity, enables the construction of more powerful and space-efficient accelerator designs.

As an example, consider the LHC main dipole magnets shown in Figure 9.4. In the LHC there are a total of 1,232 dipole magnets, each approximately 15 m long. The dipole magnets have a twin-aperture, i.e. two parallel beam pipes within each magnet. Each beam pipe is surrounded by its own set of superconducting coils each with 80 turns. Each dipole needs to generate an 8.34 T magnetic field. To achieve this magnetic field requires a current of 9.48×10^5 A in the coil, with an average current density of 300A/mm^2, which is beyond the capabilities of normal conductors.

The cables that form the superconducting coils have a rectangular cross-section to optimise packing density and ensure efficient use of space within the magnet. Where the cable consists of thousands of fine NbTi filaments embedded in

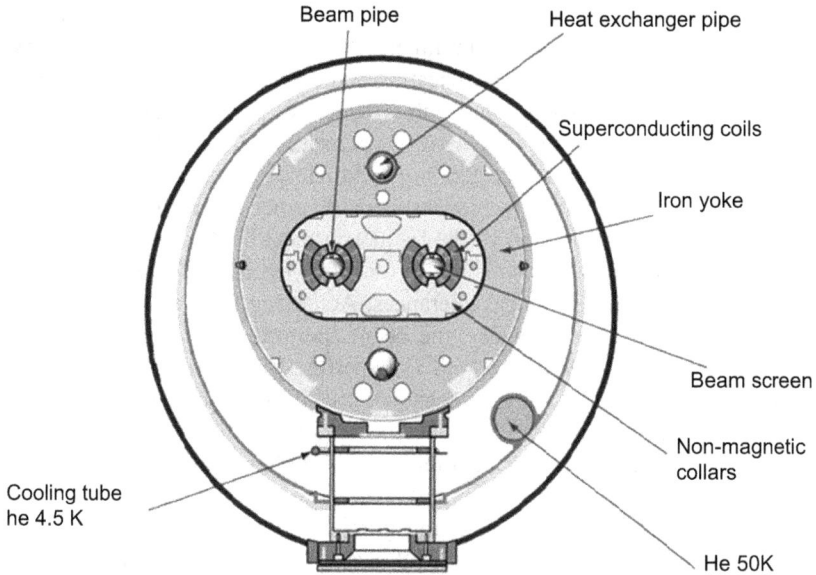

Figure 9.4 Schematic of the LHC main dipole

a copper matrix, the copper matrix provides stability and ensures uniform current distribution. Each coil consists of two layers of cables, an inner layer and an outer layer. The layers are wound in opposite directions to help balance the magnetic forces and improve field uniformity. Multiple layers of insulation, including polyimide tapes and epoxy resins, are applied to the coils to maintain, there integrity in the cryogenic environment, prevent electrical shorts and provide mechanical strength. The coils are wound in a specific pattern known as the 'cosine–theta' configuration. In this arrangement, the current density varies as the cosine of the angle around the aperture, which minimises field distortions and creates a highly uniform magnetic field in the central region of the beam pipe. The coil assembly is surrounded by an iron yoke that enhances the magnetic field strength and helps to return the magnetic flux.

9.3 Novel acceleration structures

The maximum electric field on a surface before breakdown occurs was studied empirically in 1957 by Kilpatrick, who created the Kilpatrick criterion for breakdown [20], which was later modified by Boyd for RF structures [21]

$$f = 1.64E_{pk}^2 \exp\left(-\frac{8.5}{E_{pk}}\right) \tag{9.1}$$

where E_{pk} is the maximum electric field on the surface and f is the frequency of the RF cavity. As can be seen, the maximum electric field increases with frequency. In practice, other effects such as pulsed heating and capture of field-emitted electrons reduce the steepness of the scaling with frequency. Nonetheless, the maximum electric field and hence the maximum accelerating gradient increases as we increase the frequency of the accelerating field. However, as we increase the frequency, the size of the structure reduces, causing other issues. The structures are typically synchronised together, so the frequency must be tuned within one cavity bandwidth of the accelerators synchronisation clock frequency, or a harmonic of it, and this requires tight mechanical tolerances. Achieving this becomes challenging at higher frequencies. Additionally, the beam aperture also reduces, causing the beam transport to be more challenging. The aperture will cause the beam to radiate as its image charge varies with changes in cross-section. Such effects, known as wakefields, scale up sharply with reducing beam aperture. As a compromise between these effects, the next generation of high-gradient particle accelerators will move up in frequency from the traditional S-band (3 GHz) to the X-band (11.4–12 GHz), increasing the operating gradient from around 40 up to 60–100 MV/m depending on the duty cycle [22]. Research and development is ongoing to develop X-band accelerators for high-energy physics (CLIC) [23] (shown in Figure 9.5), light sources (CompactLight) [24] and Radiotherapy (VHEE) [3]. A key goal of the research is improving our understanding of the physics of vacuum breakdown in order to increase the electric field gradient of future machines. The bulk of this research has been performed at specialist test facilities located at CERN, SLAC and KEK; however, a new generation of test facilities are being built at INFN, Cockcroft, Valencia and Melbourne.

A key focus of the research has been the question of why do we get vacuum arcs? It has long been theorised that sharp metallic whiskers on the surface of the

Figure 9.5   (Left) A cell of a CLIC accelerating structure machined into a disc, with the beam aperture and for damping waveguides shown

RF cavity emit electrons via field emission, heating the tip and resulting in vaporisation of the cavity wall material, which is then ionised creating the arc. This would appear to make sense; however, no one has ever managed to locate such a whisker on an RF cavity prior to testing and such whiskers explode into creators during the arc. Hence, the whiskers must form under RF loading. Initial testing of vacuum arc probability under DC fields has found the breakdown strength is related to the crystal structure and the stiffness of the material being tested. The RF testing has gone hand-in-hand with material science, and a key theory developed in Helsinki is that dislocations, particularly voids, under a cavity surface are stressed due to Lorentz forces under immense electric fields, giving rise to whiskers, which then arc after their formation. Such a theory would explain why the crystal structure matters and would predict the statistical nature of vacuum arcs. Further testing has supported this theory by trialling with harder or softer copper alloys showing harder alloys withstand higher fields at low breakdown rates [25] with extrapolations to breakdown rates of 1 per million pulses, suggesting a 50% improvement, although this is based on limited testing. Another key factor is the theory suggests breakdown is temperature dependant, and recent tests at SLAC have indeed shown that when cooled to cryogenic temperatures, the maximum gradient is significantly increased, with operation at 150 MV/m demonstrated to date [26], thus inspiring the Cool Copper Collider project [27].

Another key outcome is a better understanding of how the field distribution in RF cavities affects the breakdown performance. Kilpatrick's initial study had assumed that only the electric field mattered; however, when scaling RF cavities to higher frequencies (CLIC's initial design frequency was 30 GHz), it was found that at higher frequencies, there was a magnetic field dependence, thought to be related to pulsed surface heating [28]. The RF only applies heat load down to the skin depth beyond the surface, which means a very small volume. As the RF is applied in short pulses on the order of 200 ns with peak powers of tens of MW, the heating occurs on timescales significantly faster than the heat diffusion into the copper bulk, giving rise to significant surface heating on short timescales that dissipates in time as the heat diffuses. As the surface heats up while the bulk remains cold, there is a huge stress on the cavity surface that exceeds the yield strength of copper if the heating causes a temperature rise on the surface exceeding 40 K. As the pulses recur, the cyclic heating leads to surface cracking. However, when plotting the breakdown strength of several different cavity geometries, there was a large spread in both peak electric and peak magnetic fields, suggesting more complex physics was in play. In realising that the breakdown probability depends on the ability of the RF to drive large power flows into the arc location, CERN proposed that the real figure of merit would be related to the Poynting vector on the surface [29]. In travelling-wave cavities, the Poynting vector is complex, while in standing-wave cavities, the Poynting vector is purely imaginary. The potential to drive power into an arc depends on the instantaneous Poynting vector, which is smaller for the imaginary component than the real component; hence, the imaginary component is scaled in the figure of merit for breakdown, known as the modified Poynting vector. Empirical studies have shown that if the imaginary component is divided by 6 and added to the real part, then most

RF structures break down at a very similar modified Poynting vector when scaled to the same pulse length, structure length, and breakdown rate.

Recent studies have started to look at increasing the frequency to Ka-band (30–36 GHz) as RF linearisers, which remove curvature from the beams' phase-space due to the sinusoidal variation of the RF with time in the lower frequency injectors. The lineariser requires a time-varying voltage to straighten out the RF field, and hence, a high change in voltage with time is needed. As a result, the voltage required for linearisation decreases roughly proportional to the square of the lineariser's frequency [30]. In addition, for light sources that operate with single bunches as opposed to a train, the pulse duration is only limited by the fill time of the cavity, which is frequency dependent, hence further increasing the maximum gradient. One of the main deficiencies with Ka-band operation is the limited peak power available from Ka-band amplifiers; work is ongoing to develop HOM Klystrons, Gyro-klystrons and Upconverters to fill this gap.

To create a step change in RF gradient and time structures, there has been recent interest in further frequency increases to the millimetre wave and THz (100–400 GHz) range [31]. Such structures require completely different manu-facturing methods to achieve the requisite miniaturisation and the sub-mm-sized aperture requires new types of beam transport. 200 GHz metallic structures have been tested with full breakdown statistics at SLAC but to date have not exceeded the gradients obtained at lower frequencies, achieving 56 MV/m [32]. Another advantage of the higher frequencies is that as the pulse lengths are proportionally shorter, it is possible to obtain very high peak powers while keeping the pulse energy low. Currently, laser-based THz sources can generate peak powers in excess of a GW. In addition to potential increases in gradient up to several hundred MV/m, the small RF periods offer the ability to create, manipulate and compress bunches to scales not possible with RF accelerators, while having solid structures allows these accelerators to operate with good pulse-to-pulse stability and excellent beam properties when compared to plasma accelerators. In order to drive these structures, there are three options being considered: Gyrotron oscillators and amplifiers, non-linear crystals using lasers and beam-driven approaches. Gyrotrons are a mature technology with large powers being developed for plasma heating at 100 GHz; however, they are long pulse devices, requiring the pulse to be chopped, making them very inefficient, although in future short pulse Gyro-TWA's and Gyro-Klystrons may be developed with 100 GHz Gyro-TWA's already delivering 100 kW and designs for 10 MW Gyro-Klystrons in the literature [33]. Laser-based sources are based on the harmonics produced and hence are already inefficient but can deliver up to GW of peak power in pulses of a few picoseconds. As the filling time is much shorter at high frequency, there are efficiency savings; hence, the source can tolerate being a little more inefficient than RF sources. Beam-driven structures have already demonstrated gradients well over 100 MV/m with some evidence of GV/m gradients; however, the large bunch charges required for the drive beam excite wakefields in the structures, which can disrupt the witness beam requiring a separate accelerator to create the drive beam [34]. There is also interest in direct laser acceleration extending the frequencies into the infrared wavelength, but with such scaled-down apertures and wavelengths, getting bunches of appreciable charge through the structure is a considerable challenge [35].

The structures proposed for the mm-wave and THz regimes are typically of three types.

- Metallic coupled-cavity or corrugated waveguide structures that resemble miniature RF accelerators. The challenge here is the manufacture of the corrugations, which limits the frequency to just above 100 GHz.
- Dielectric-lined waveguides which are circular or rectangular waveguides with a layer of dielectric on the inside of the walls, slowing down the electromagnetic waves' phase velocity to match the beam velocity. This requires some glue to hold the dielectric in place, but such structures have been demonstrated up to 400 GHz, as shown in Figure 9.6.
- Photonic structures: typically all-dielectric which avoids the need for the metallic wall which can be lossy at higher frequencies. While the structures are typically more complex, they are actually simpler to make at high frequencies due to recent advances in manufacturing small-scale dielectrics. Photonic structures have been demonstrated at optical frequencies but not at high gradients with a beam present [35].

Other acceleration mechanisms exist, such as direct THz acceleration using a travelling wave source with a tilted pulse front [36] and inverse free-electron lasers [37], but these are less common.

Work is now beginning on scaling these structures to full-scale accelerators. There are several challenges that still need to be addressed. One such challenge is

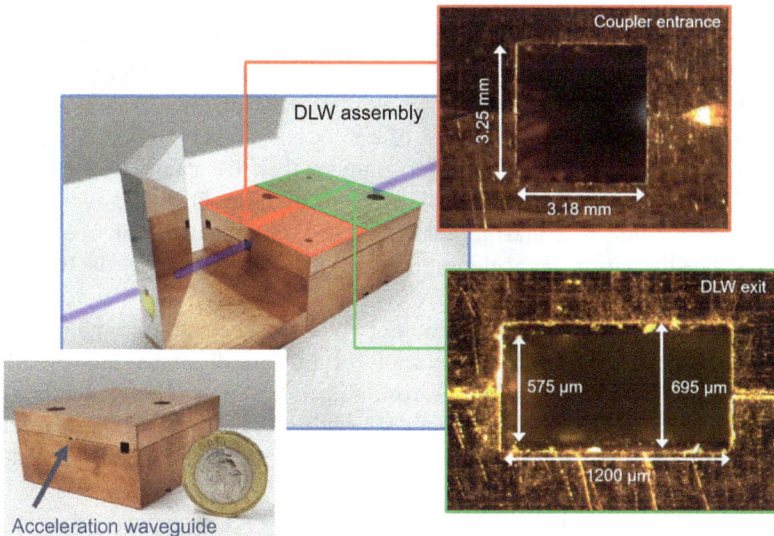

Figure 9.6   (Left) 400 GHz dielectric lined waveguide

beam transport with small apertures [38]. One proposal is to use a FODO lattice with the structures placed inside the magnet apertures at the beam size minima, taking advantage of the small size of these accelerators. For rectangular geometries, the structures can be interleaved with horizontal and vertical orientations such that the structures' quadrupole field components cancel each other out of every two structures. As the charge increases, we get beam stability problems associated with the transverse wakefield and energy depletion as the energy transferred to the beam becomes comparable in scale to the energy in the accelerating wave. For these reasons mm-wave accelerators are unlikely to be able to rival the highest beam charges that can be delivered by RF accelerators. To compensate for this, due to the small size of the structures, it could be possible to have several structures in parallel, each taking part of the total charge, analogous to a multi-beam klystron, which can then be combined at the end of the accelerator.

Not all applications require mm-wave and THz structures to be the main accelerating workforce, as these structures can also play a role in the machine by their use in beam manipulation or diagnostics. Work at SLAC has shown that a THz accelerator placed between the electron gun and the main accelerator can improve beam arrival time and energy jitter by synchonising the THz to have the ideal bunch arrive at the zero crossing: hence earlier or late bunches get an equal and opposite acceleration/deceleration. THz deflecting structures have also been used to produce a streaking of picosecond-scale electron bunches, giving a bunch longitudinal profile with a time resolution significantly beyond RF deflectors.

9.4　Plasma accelerators

Radiofrequency (RF) accelerators are limited to accelerating gradients of <100 MV/m. This implies that to reach higher particle energies, it is necessary to have longer accelerators. For example, the International Linear Collider is a proposed 500 GeV electron–positron collider approximately 30 km long. A large part of the expense of such a machine is the cost of boring tens of kilometres of tunnels; an accelerating technology with a higher gradient than RF cavities would clearly reduce the total length, and thus cost, of the accelerator. One such technology is particle acceleration in a plasma, that is, a medium consisting of free electrons and ions. The limitation of acceleration in an RF cavity is given by the breakdown of the material of the cavity itself. By contrast, plasma cannot be further broken down so is not susceptible to this limitation. Plasma accelerators can be driven by either a powerful laser or another particle beam. In both cases, the drive beam excites an oscillation of the free electrons in the plasma. This generates large charge separation, and thus electric fields, within the plasma, which can accelerate electrons either from within the plasma itself or injected from an external source.

For laser plasma wakefield acceleration (LWFA), a typical experimental arrangement involves focusing a high-power (TW or PW) pulsed laser into a gas, where the leading edge of the laser pulse ionises the gas, creating the plasma.

Electrons are expelled from the high-intensity region of the laser pulse by the pondermotive force and create large amplitude oscillations (wakefields) that travel at the speed of the laser pulse [39]. Typically, these experiments are carried out in the nonlinear or 'wave-breaking' regime, where background electrons in the plasma can be trapped and accelerated in the large electric fields created in the plasma. These can be many orders of magnitude larger than in conventional RF accelerators. Experiments have demonstrated the acceleration of electrons to 3.25 GeV in 14 mm (>200 GV/m) [40] and 8 GeV in 200 mm (40 GV/m) [41]. This technology has massive potential for reducing the size and, thus, cost of particle accelerators. However, to be competitive, LWFA needs to demonstrate the same high levels of repeatability, stability, beam charge and quality as conventional RF accelerators. Considerable progress has been made in this area, including the demonstration of sub-percent energy spread electron beams [42], <1 mm mrad emittance beams [43], high charge of up to 500 pC [44], and ultrashort, few femtosecond electron bunches [45]. Machine learning techniques have been successfully applied to LWFA, showing the potential for optimising the output of these experiments [46] and the possibility of using machine learning and feedback to control a plasma accelerator. There have also been impressive results shown in the area of long-term operation of laser-driven plasma accelerators, with the demonstration of stable operation over 24 h and 100,000 consecutive shots [47]. This is a very important step in the move from laser-plasma accelerators being research devices to machines capable of operating reliably for long periods.

Another important demonstration of the quality of beams from LWFA is their use to drive free electron lasers (FELs, such as the Linac Coherent Light Source-II at the SLAC National Accelerator Laboratory in the US or the European X-ray FEL in Germany). FELs are important sources of X-rays, which are generated by electron bunches passing through an alternating structure of magnets called an undulator, and they have demanding requirements for the charge, stability, energy and emittance of the electron bunches used to drive them [48]. FELs are national-scale facilities and typically require conventional electron accelerators several kilometres in length. It has now been shown that an FEL can be driven by electron bunches from laser-driven plasma accelerators [49], where the electron acceleration region in the plasma was only 6 mm long, and the total beamline, including magnets and undulator 12 m. Lasing of an FEL has also been demonstrated from a compact particle beam-driven plasma accelerator [50]. This again demonstrates the huge potential for plasma acceleration to significantly reduce the size and cost of particle accelerators and machines, such as FELs, that rely on accelerators and thus increase their availability to users.

However, there are still a number of challenges facing LWFA before it can be considered a viable alternative to conventional RF acceleration. These include the laser driver itself, the plasma source, the staging of multiple accelerating regions, and the quality of the output electron bunches.

Laser: Typically, the drivers for LWFA are titanium sapphire (Ti:sapp) lasers, which can produce pulses with several tens of Joules of energy in <100 fs and hence peak powers in the TW or even PW range [51]. These systems have been hugely successful in research but have some problems as drivers of practical machines. The first is the repetition rate of the pulses, which can range from 10 Hz to one pulse every several minutes. Ideally, systems should be capable of running at several kilohertz to allow data collection at the rates required by, for example, particle physics experiments and for feedback mechanisms to be successfully implemented on the accelerator. Second, high-peak-power lasers typically have very low wall plug efficiency (<0.1 percent). Taking into consideration an efficiency of energy transfer from the laser to the plasma of 50 percent [52] and from the plasma to the electron beam of 20–40 percent [53,54], it can be seen that overall LWFAs have low efficiency and that the main driver of this inefficiency is the laser. It is clear that for a practical and sustainable accelerator to be built using LWFA technology lasers will have to be developed that are both high peak but also high average power, i.e. with high repetition rates, and that these lasers will need to have considerably higher wall plug efficiencies, in the several tens of percent, than currently available. A number of options are under consideration by researchers worldwide to improve laser drivers for LWFA. The first approach is to improve current Ti:sapp lasers in terms of efficiency, but especially repetition rate. Projects aim to improve both pump lasers and cooling of the laser medium to achieve repetition rates of 10–100 Hz, an order of magnitude improvement on current operation [55,56], sufficient to drive the next generation of LWFA experiments investigating stability and long-term 24/7 operation. In the long term, it is clear that Ti:sapp lasers will not be efficient enough for the operation of a future facility where sustainability will be a major driver of the choice of technology. For this application, a different approach to the laser driver will be needed. Current research covers two main areas – utilising a different laser medium [57], or taking smaller, more efficient fibre lasers and coherently combining them together to generate the high-peak-power pulses required for LWFA [58].

Injection: the source of electrons in an LWFA is usually the background plasma itself. When the wake is driven sufficiently strongly (the nonlinear or 'bubble' regime), some electrons are trapped and accelerated behind the laser pulse. This has been very successful in producing high-energy electron bunches but is an uncontrolled process and contributes to the shot-to-shot instability and energy spread of LWFAs. Improving the output beam quality of these accelerators requires either more controlled injection within the plasma (e.g. separate ionisation of a different gas species within a mixture [59], longitudinally varying the plasma density [60]), or the injection of electrons from a separate external source, i.e. an RF linac [61].

Plasma sources: a plasma accelerator requires the ionisation of neutral gas within a vacuum chamber. The gas target can be of several different types: a jet, where the gas comes out of a small nozzle opening backed by a high-pressure gas reservoir, a cell, a solid structure containing the gas that has holes or windows for the laser to enter, and a capillary, a narrow cylindrical hole several hundred microns in diameter drilled into a hard substance such as sapphire, with gas pumped in through one or more holes longitudinally where the gas is typically ionised by an electrical discharge rather than the driving laser pulse. These have all been used successfully in LWFA experiments but capillaries and cells both have solid material very close to the laser beam, whereas gas jets can be several mm away. This means that capillaries and cells are vulnerable to laser damage if there is any transverse movement of the laser focus (spatial jitter) and therefore have limited lifetimes before they need to be replaced. LWFAs operating at kHz repetition rates will require sources that last billions of shots, and gas targets need to be developed that can be used under these conditions.

Guiding in plasma: a tightly focused laser beam, as required to produce the high intensities for LWFA, will diffract over a short distance. This means the length over which particles are accelerated is short, so even for a large accelerating gradient the total energy gain, given by the gradient multiplied by the length, will be small. It is necessary to find a mechanism to guide the high-intensity laser beam over several centimetres, or even metres, to achieve energy gains of many GeV. For low-intensity lasers, such a waveguide is an optical fibre, but these would be destroyed by a high-intensity pulse. The solution is to generate a 'plasma optical fibre'. A hot column of electrons is created in the plasma, either by electrical discharge (for a capillary system) or using a preliminary laser pulse focused on a line using special optics [62]. The column expands radially, creating a structure in the plasma with a lower density of electrons on axis surrounded by a higher density annulus. As the refractive index in a plasma increases with decreasing electron density, this creates a guiding structure analogous to an optical fibre within the plasma. These plasma waveguides have been shown to guide high-intensity pulses over many centimetres, allowing energy gains of up to 8GeV in a capillary [41] and 5GeV in a gas jet [63]. These successful demonstrations will need to be scaled to longer distances of up to a metre to achieve higher energy electrons.

Staging: as well as diffraction, LWFAs are subject to dephasing and depletion [39]. Dephasing is due to the different velocities of the laser (and therefore the wakefield), and the electrons in the plasma. As the velocity of the electrons approaches c, they outrun the laser pulse and can enter a decelerating phase of the wake. This effect can be reduced by using a lower density plasma. The final limitation is depletion, where so much of the energy in the laser pulse has been transferred to the plasma that the laser can no longer efficiently drive a wakefield. Given that diffraction can be overcome using the guiding techniques outlined above, and dephasing by using low plasma densities, the fundamental limit on the length, and hence electron energy, achievable in one plasma acceleration stage is set by the depletion of the laser pulse. As a single laser pulse does not have the energy to accelerate pC of charge to hundreds of GeV or even TeV, it follows that to generate very high particles, e.g. for a particle physics linear collider, it is necessary to combine multiple accelerating stages together. Research in this area is concentrated on the challenges of extracting and capturing an accelerated bunch from one plasma stage and transporting it to another without losing charge or degrading the emittance of the beam [64]. This requires significant research into magnet technology as well as the plasma accelerator itself.

Plasma accelerators may also be driven by a charged particle beam (plasma wakefield acceleration or PWFA). This has been demonstrated using both electron and proton beams as the drivers [65,66]. PWFAs share many similarities with LWFA systems, but typically the driver beam, either electron or proton, will not be able to ionise a neutral gas so PWFAs require a method of generating plasma, usually a laser or electrical discharge, before the arrival of the drive beam which creates the accelerating wake. The driver can be shorter than the plasma period and drive a wake resonantly with a single bunch, usual for electron-driven experiments, or in the case of the AWAKE project, a long proton bunch from the CERN Super Proton Synchrotron is used as the driver. This is significantly longer than the period even for a low-density plasma, and would therefore not be expected to drive a wakefield efficiently. However, through a process of self-modulation in the plasma, the initially uniform proton beam forms into resonantly spaced micro-bunches that have been shown to successfully drive high accelerating gradients [65]. PWFAs are not typically driven in the highly nonlinear bubble regime of LWFAs where wavebreaking directly injects background into the wake to be accelerated. Therefore, they require the external injection of electrons. In early electron-drive experiments, the electrons at the back of the drive bunch were accelerated in the wake created by the front of the bunch, demonstrating high energy gain [66], but more recent experiments have concentrated on the production of separate drive and witness beams, where the delay between the two bunches can be optimised so the witness is injected into the accelerating phase of the wakefield [67]. To avoid the complexity of generating multiple bunches in an RF linac, alternative approaches using laser ionisation within the plasma to release electrons directly into the wake have also been demonstrated [68], with the aim of producing ultra-highbrightness beams. For the AWAKE project, a separate linac has been used to inject electrons into the plasma. PWFAs have the advantage over LWFAs of having higher energy drive beams (typically 10s to 100s kJ in the case of AWAKE), which can be transferred to the witness bunch, and not require guiding over acceleration distances of several metres. They also do not suffer from dephasing, but of course require the RF infrastructure to generate the drive beam in the first place, thus potentially negating the major advantage of plasma accelerators, i.e. a small footprint.

Simulations: theory and simulations are a critical part of the development of plasma accelerators [69], especially as there are limited facilities for experimentation worldwide and access is restricted. Therefore, numerical experiments are vital for identifying the best parameter space for experiments and understanding the challenges of plasma acceleration. The main tool for simulations of drive beam/ plasma interactions are particle-in-cell (PIC) codes. These have reached a high level of sophistication, but the need to resolve physical processes at the tens of nanometre scale over acceleration distances of many centimetres means they are very computationally expensive. Significant work is ongoing to speed up these models by the use of boosted frames and, where appropriate, approximations such as reduced dimensionality.

It is not currently clear if plasma-based accelerators will become a competitive technology for future high-energy physics colliders, but they certainly have a promising future in lower energy applications such as light sources and FELs.

European and US roadmaps for accelerator development recommend further research and development into plasma acceleration and laser development [70,71] as a route to a sustainable, affordable collider, and it is expected that this remains a vibrant field of research for the next decades.

References

[1] Applications of Particle Accelerators in Europe. *Technical report*, CERN, Geneva, 2020.

[2] C. Young, M. Chen, T. Chang, C. Ko, and K. Jen. Cascade Cockcroft–Walton voltage multiplier applied to transformerless high step-up dc–dc converter. *IEEE Transactions on Industrial Electronics*, 60(2):523–537, 2013.

[3] L. Whitmore, R. I. Mackay, M. Van Herk, J. K. Jones, and R. M. Jones. Focused vhee (very high energy electron) beams and dose delivery for radiotherapy applications. *Scientific Reports*, 11(1):14013, 2021.

[4] C. M. Charlie Ma and T. Lomax. *Proton and Carbon Ion Therapy*. Boca Raton, FL: CRC Press, 2012.

[5] D. Kramer. Beating back the coronavirus requires a bigger arsenal. *Physics Today*, 74 (4), 20–23, 2021.

[6] W. Wuensch. Advances in the understanding of the physical processes of vacuum breakdown. *High Gradient Accelerating Structure*, pp. 31–50, 2014.

[7] D. Boussard and T. P. R. Linnecar. The LHC superconducting RF system. *Technical report*, 1999.

[8] H. Padamsee, J. Knobloch, and T. Hays. *RF Superconductivity for Accelerators*. New York: Wiley, 2008.

[9] H. K. Onnes. The superconductivity of mercury. *Communications from Physical Laboratory of the University of Leiden*, pp. 122–124, 1911.

[10] L. N. Cooper. Bound electron pairs in a degenerate fermi gas. *Physical Review*, 104:1189–1190, 1956.

[11] J. Bardeen and D. Pines. Electron–phonon interaction in metals. *Physical Review*, 99:1140–1150, 1955.

[12] J. Bardeen, L. N. Cooper, and J. R. Schrieffer. Microscopic theory of superconductivity. *Physical Review*, 106:162–164, 1957.

[13] J. Bardeen, L. N. Cooper, and J. R. Schrieffer. Theory of superconductivity. *Physical Review*, 108:1175–1204, 1957.

[14] J. Matricon and D. Saint-James. Superheating fields in superconductors. *Physical Letters*, 24A:241–242, 1967.

[15] B. Aune, R. Bandelmann, D. Bloess, *et al.* Superconducting tesla cavities. *Physical Review ST Accelerators and Beams*, 3:092001, 2000.

[16] M. Ghaemi, A. Lopez-Cazalilla, K. Sarakinos, *et al.* Growth of nb films on cu for superconducting radio frequency cavities by direct current and high power impulse magnetron sputtering: A molecular dynamics and experimental study. *Surface and Coatings Technology*, 476:130199, 2024.

[17]   S. A. Willson, R. G. Farber, A. C. Hire, R. G. Hennig, and S. J. Sibener. Submonolayer and monolayer Sn adsorption and diffusion behavior on oxidized Nb(100). *Journal of Physical Chemistry C*, 127(6):3339–3348, 2023.

[18]   A. M. Valente-Feliciano, C. Antoine, S. Anlage, *et al.* Next-generation superconducting RF technology based on advanced thin film technologies and innovative materials for accelerator enhanced performance and energy reach, 2022.

[19]   E. Barzi, M. Bestetti, F. Reginato, D. Turrioni, and S. Franz. Synthesis of superconducting nb3sn coatings on nb substrates. *Superconductor Science and Technology*, 29(1):015009, 2015.

[20]   W. D. Kilpatrick. Criterion for vacuum sparking designed to include both RF and DC. *Review of Scientific Instruments*, 28(10):824–826, 1957.

[21]   T. J. Boyd Jr. Kilpatrick's criterion. *Los Alamos Group AT-1 Report*, 82:28, 1982.

[22]   A. Degiovanni, W. Wuensch, and J. G. Navarro. Comparison of the conditioning of high gradient accelerating structures. *Physical Review Accelerators and Beams*, 19:032001, 2016.

[23]   H. Zha and A. Grudiev. New CLIC-G structure design. *Technical report*, 2016.

[24]   G. d'Auria, A. W. Cross, L. Nix, *et al.* The compact light design study. In *10th International Particle Accelerator Conference*, IPAC 2019, pp. 1756–1759, 2019.

[25]   V. A. Dolgashev, L. Faillace, B. Spataro, S. Tantawi, and R. Bonifazi. High-gradient RF tests of welded X-band accelerating cavities. *Physical Review Accelerators and Beams*, 24(8):081002, 2021.

[26]   M. Nasr, E. Nanni, M. Breidenbach, S. Weathersby, M. Oriunno, and S. Tantawi. Experimental demonstration of particle acceleration with normal conducting accelerating structure at cryogenic temperature. *Physical Review Accelerators and Beams*, 24(9):093201, 2021.

[27]   C. Vernieri, E. A. Nanni, S. Dasu, *et al.* A "cool" route to the Higgs boson and beyond the cool copper collider. *Journal of Instrumentation*, 18(07): P07053, 2023.

[28]   D.P. Pritzkau and R.H. Siemann. Experimental study of RF pulsed heating on oxygen free electronic copper. *Physical Review ST Accelerators and Beams*, 5(11):112002, 2002.

[29]   A. Grudiev, S. Calatroni, and W. Wuensch. New local field quantity describing the high gradient limit of accelerating structures. *Physical Review ST Accelerators and Beams*, 12:102001, 2009.

[30]   A. Castilla, R. Apsimon, G. Burt, *et al.* Ka-band linearizer structure studies for a compact light source. *Physical Review Accelerators and Beams*, 25(11): 112001, 2022.

[31]   A. L. Lake D. S. Georgiadis V Smith, *et al.* Acceleration of relativistic beams using laser-generated terahertz pulses. *Nature Photonics*, 14(12): 755–759, 2020.

[32] M. D. Forno, V. Dolgashev, G. Bowden, *et al.* RF breakdown measurements in electron beam driven 200 GHz copper and copper–silver accelerating structures. *Physical Review Accelerators and Beams*, 19 (11):111301, 2016.

[33] W. Lawson, R. Lawrence Ives, M. Mizuhara, J. M. Neilson, and M. E. Read. Design of a 10-MW, 91.4-GHz frequency-doubling gyroklystron for advanced accelerator applications. *IEEE Transactions on Plasma Science*, 29 (3):545–558, 2001.

[34] B. D. O'Shea, G. Andonian, S. K. Barber, *et al.* Observation of acceleration and deceleration in gigaelectron-volt-per-metre gradient dielectric wakefield accelerators. *Nature Communications*, 7:12763, 2016.

[35] J. R. England, R. J. Noble, K. Bane, *et al.* Dielectric laser accelerators. *Reviews of Modern Physics*, 86(4):1337, 2014.

[36] D. A. Walsh, D. S. Lake, E. W. Snedden, M. J. Cliffe, D. M. Graham, and S. P. Jamison. Demonstration of sub-luminal propagation of single-cycle terahertz pulses for particle acceleration. *Nat. Commun.*, 8(1):421, 2017.

[37] E. Curry, S. Fabbri, J. Maxson, P. Musumeci, and A. Gover. Meter-scale terahertz-driven acceleration of a relativistic beam. *Phys. Rev. Letters*, 120 (9):094801, 2018.

[38] Ö. Apsimon, G. Burt, R. B. Appleby, R. J. Apsimon, D. M. Graham, and S. P. Jamison. Six-dimensional phase space preservation in a terahertz-driven multistage dielectric-lined rectangular waveguide accelerator. *Physical Review Accelerators and Beams*, 24(12):121303, 2021.

[39] E. Esarey, C. B. Schroeder, and W. P. Leemans. Physics of laser-driven plasma-based electron accelerators. *Rev. Mod. Phys.*, 81:1229–1285, 2009.

[40] H. T. Kim, K. H. Pae, H. J. Cha, *et al.* Enhancement of electron energy to the multi-GeV regime by a dual-stage laser-wakefield accelerator pumped by petawatt laser pulses. *Physical Review Letters*, 111:165002, Oct 2013.

[41] A. J. Gonsalves, K. Nakamura, J. Daniels, *et al.* Petawatt laser guiding and electron beam acceleration to 8 GeV in a laser-heated capillary discharge waveguide. *Physical Review Letters*, 122:084801, 2019.

[42] W. T. Wang, W. T. Li, J. S. Liu, *et al.* High-brightness high-energy electron beams from a laser wakefield accelerator via energy chirp control. *Physical Review Letters*, 117:124801, 2016.

[43] E. Brunetti, R. P. Shanks, G. G. Manahan, *et al.* Low emittance, high brilliance relativistic electron beams from a laser-plasma accelerator. *Physical Review Letters*, 105:215007, 2010.

[44] Y. F. Li, D. Z. Li, K. Huang, *et al.* Generation of 20 kA electron beam from a laser wakefield accelerator. *Physics of Plasmas*, 24(2):023108, 2017.

[45] O. Lundh, J. Lim, C. Rechatin, *et al.* Few femtosecond, few kiloampere electron bunch produced by a laser–plasma accelerator. *Nature Physics*, 7:219, 2011.

[46] R. J. Shalloo, S. J. D. Dann, J.-N. Gruse, *et al.* Automation and control of laser wakefield accelerators using bayesian optimization. *Nature Communications*, 11(1):6355, 2020.

[47] A. R. Maier, N. M. Delbos, T. Eichner, *et al.* Decoding sources of energy variability in a laser-plasma accelerator. *Physical Review X*, 10:031039, 2020.

[48] N. Huang, H. Deng, B. Liu, D. Wang, and Z. Zhao. Features and futures of X-ray free-electron lasers. *The Innovation*, 2(2):100097, 2021.

[49] W. Wang, K. Feng, L. Ke, *et al.* Free-electron lasing at 27 nanometres based on a laser wakefield accelerator. *Nature*, 595(7868):516–520, 2021.

[50] R. Pompili, D. Alesini, M. P. Anania, *et al.* Free-electron lasing with compact beam-driven plasma wakefield accelerator. *Nature*, 605(7911):659–662, 2022.

[51] C. N. Danson, C. Haefner, J. Bromage, *et al.* Petawatt and exawatt class lasers worldwide. *High Power Laser Science and Engineering*, 7:e54, 2019.

[52] B. A. Shadwick, C. B. Schroeder, and E. Esarey. Nonlinear laser energy depletion in laser-plasma acceleratorsa). *Physics of Plasmas*, 16(5):056704, 2009.

[53] T. C. Katsouleas, J. J. Su, S. C. Wilks, J. M. Dawson, and P. Chen. Beam loading in plasma accelerators. *Particle Accelerators*, 22:81–99, 1987.

[54] M. Tzoufras, W. Lu, F. S. Tsung, *et al.* Beam loading in the nonlinear regime of plasma-based acceleration. *Physical Review Letters*, 101:145002, 2008.

[55] R. W. Assmann, M. K. Weikum, T. Akhter, *et al.* Eupraxia conceptual design report. *The European Physical Journal Special Topics*, 229(24):3675–4284, 2020.

[56] P. Mason, M. Divoký, K. Ertel, *et al.* Kilowatt average power 100 J-level diode pumped solid state laser. *Optica*, 4(4):438–439, 2017.

[57] I. Tamer, Z. Hubka, L. Kiani, *et al.* Demonstration of a 1 TW peak power, joule-level ultrashort Tm:YLF laser. *Optical Letters*, 49(6):1583–1586, 2024.

[58] A. Klenke, M. Müller, H. Stark, *et al.* Coherent beam combination of ultrafast fiber lasers. *IEEE Journal of Selected Topics in Quantum Electronics*, 24 (5):1–9, 2018.

[59] E. Oz, S. Deng, T. Katsouleas, *et al.* Ionization-induced electron trapping in ultrarelativistic plasma wakes. *Physical Review Letters*, 98:084801, 2007.

[60] C. G. R. Geddes, K. Nakamura, G. R. Plateau, *et al.* Plasma-density-gradient injection of low absolute-momentum-spread electron bunches. *Physical Review Letters*, 100:215004, 2008.

[61] Y. Wu, J. Hua, Z. Zhou, *et al.* High-throughput injection – acceleration of electron bunches from a linear accelerator to a laser wakefield accelerator. *Nature Physics*, 17(7):801–806, 2021.

[62] R. J. Shalloo, C. Arran, L. Corner, *et al.* Hydrodynamic optical-field-ionized plasma channels. *Physical Review E*, 97:053203, 2018.

[63] J. E. Shrock, B. Miao, L. Feder, and H. M. Milchberg. Meter-scale plasma waveguides for multi-GeV laser wakefield acceleration. *Physics of Plasmas*, 29(7):073101, 2022.

[64] S. Steinke, J. van Tilborg, C. Benedetti, *et al.* Staging of laser-plasma accelerators. *Physics of Plasmas*, 23(5):056705, 2016.

[65] E. Adli, A. Ahuja, O. Apsimon, *et al.* Acceleration of electrons in the plasma wakefield of a proton bunch. *Nature*, 561(7723):363–367, 2018.

[66] I. Blumenfeld, C. E. Clayton, F.-J. Decker, *et al.* Energy doubling of 42 GeV electrons in a metre-scale plasma wakefield accelerator. *Nature*, 445 (7129):741–744, 2007.

[67] R. Pompili, D. Alesini, M. P. Anania, *et al.* Energy spread minimization in a beam-driven plasma wakefield accelerator. *Nature Physics*, 17(4):499–503, 2021.

[68] A. Deng, O. S. Karger, T. Heinemann, *et al.* Generation and acceleration of electron bunches from a plasma photocathode. *Nature Physics*, 15(11):1156–1160, 2019.

[69] J.-L. Vay and R. Lehe. Simulations for plasma and laser acceleration. *Reviews of Accelerator Science and Technology*, 9:165–186, 2016.

[70] C. Adolphsen, D. Angal-Kalinin, T. Arndt, *et al.* European strategy for particle physics – accelerator R&D roadmap. *European Strategy for Particle Physics – Accelerator R D Roadmap, N. Mounet (ed.), CERN Yellow Reports: Monographs, CERN-2022-001 (CERN, Geneva, 2022)*, 2022.

[71] S. Asai, A. Ballarino, T. Bose, *et al.* Report of the 2023 particle physics project prioritization panel. 2023.

Chapter 10
Prospective technological applications of magnetic monopoles
Andrew Michael Chugg[1]

10.1 Missing magnetic monopoles

The concept of electrical charge is a fundamental pillar of physics. Recorded observations of static electrical charging effects date back at least as far as the ancient Greeks. Specifically, around 600 BC Thales of Miletus observed that amber, when rubbed with animal fur, can attract feathers and similar lightweight materials. In ancient Greek writings that record Thales' work and theories, amber is called *electron*, and this term has provided the derivation for the word electricity as used today. In modern physics, the charge is assigned to fundamental particles, notably electrons and protons. It is particularly the flow of these charged particles, especially electrons, through materials, which is referred to as electricity. As is well known and appreciated, the phenomenon of electricity energizes and drives a large proportion of the technological apparatus currently used by humans.

However, Thales also investigated the property of a mineral called lodestone to attract iron. It is suggested that lodestones were found in Magnesia and therefore became referred to as magnets. Nowadays, we call the lodestone ore "magnetite" and its attraction for iron is recognized as an example of magnetism. In the nineteenth century, through the work of Michael Faraday and James Clerk Maxwell, the first great unification of the fundamental forces of nature was achieved by showing that electricity and magnetism are simply variant manifestations of the same underlying electromagnetic force. This unification was enshrined in a set of four mathematical relationships between the electric field E, the magnetic field H, the electrical charge density ρ_e, and the electrical current density J_e. These relationships include a parameter called the permittivity ε that describes the strength of electric fields set up by charges and a parameter called the permeability μ that describes the strength of magnetic fields set up by electrical currents. These four formulae are known collectively as the Maxwell equations.

[1]MBDA UK Limited, Bristol, UK

The first Maxwell equation is Gauss's law for electric fields:

$$\nabla . E = \rho_e / \varepsilon \tag{10.1}$$

This law defines electric charge as the mathematical divergence of the electric field and thereby designates electric charge as a source of or sink for electric field lines depending on its sign.

The second Maxwell equation is Gauss's law for magnetic fields:

$$\nabla . H = 0 \tag{10.2}$$

This law defines the divergence of the magnetic field to be zero everywhere, which is the same as defining there to be no sources of or sinks for magnetic field lines anywhere in the universe. In other words, all magnetic field lines loop around and bite their own tails like the Ouroboros snake ring.

The third Maxwell equation is Faraday's law of electromagnetic induction:

$$\nabla \times E = -\mu \frac{\partial H}{\partial t} \tag{10.3}$$

This law defines that a rate of change of a magnetic field induces an associated electric field, as demonstrated by Faraday's experiments with magnets.

The fourth and final Maxwell equation is a version of Ampère's law for sourcing a magnetic field from an electrical current density, but Maxwell extended it with the addition of a term indicating that a rate of variation of an electric field also sources an associated magnetic field:

$$\nabla \times H = -\varepsilon \frac{\partial E}{\partial t} + J_e \tag{10.4}$$

This law defines two options for producing a magnetic field: either by setting up an electrical current or by varying the magnitude and direction of an electric field.

The Maxwell equations are among the most perfect and inviolable laws in all of physics. For example, although they were formulated before Einstein had defined his special and general theories of relativity, the Maxwell equations turned out already to accommodate the changes that Einstein implemented for the laws of mechanics. Neither is there any electrical, magnetic, or electromagnetic phenomenon known to physics that is not precisely explained by Maxwell's equations. This includes electromagnetic waves since it transpires that a pair of interlinked or coupled wave equations for the electromagnetic fields are directly derivable from the Maxwell equations:

$$\nabla^2 E = \mu\varepsilon \frac{\partial^2 E}{\partial t^2} \tag{10.5}$$

$$\nabla^2 H = \mu\varepsilon \frac{\partial^2 H}{\partial t^2} \tag{10.6}$$

These equations predicted the existence of electromagnetic waves before they had been discovered, and they additionally define the propagation velocity v of electromagnetic waves:

$$v = \frac{1}{\sqrt{\mu\varepsilon}} = \frac{1}{\sqrt{\mu_r\varepsilon_r\mu_0\varepsilon_0}} = \frac{c}{\sqrt{\mu_r\varepsilon_r}} \tag{10.7}$$

where the permeability and permittivity are split into their respective values in vacuum, μ_0 and ε_0, and dimensionless coefficients, μ_r and ε_r, which adapt their values to those observed within a particular medium. It quickly became evident that v in vacuum was equal to c, the speed of light in vacuum, which was a known property of light. This implied that light could be a type of electromagnetic wave. The reality of electromagnetic waves was dramatically confirmed by Heinrich Hertz in experiments between 1886 and 1889 using a crude apparatus in which a spark was created across a small gap by discharging a pair of capacitor plates, and this was shown to induce a spark in the spark gap of a receiver loop on the far side of the laboratory (Figure 10.1). When queried at the time, Hertz failed to specify any useful application for his discovery, yet within a decade, Marconi was transmitting signals over vast distances at the speed of light using radio waves.

However, there is another even clearer and more dramatic prediction innate within the form of the Maxwell equations. This is the existence of magnetic charges paralleling and complementing the electrical charges with which we are now so familiar. There is an almost inexorable symmetry argument that demands that the second Maxwell equation be amended with the existence of magnetic charge

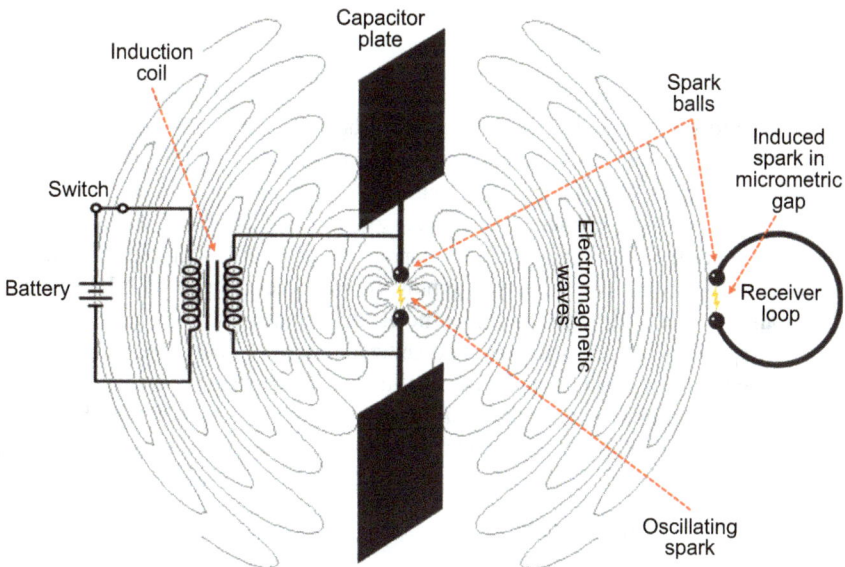

Figure 10.1   The first transmission and detection of electromagnetic waves by Heinrich Hertz

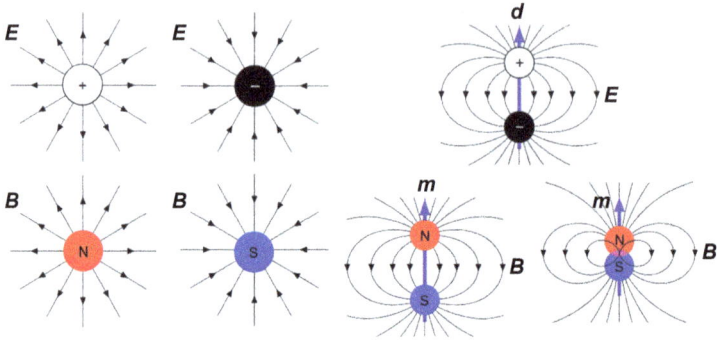

Figure 10.2   Field line configurations for electric and magnetic monopoles and electric and magnetic dipoles (Maschen, Public Domain: https:// commons.wikimedia.org/wiki/File:Em_monopoles.svg and https:// commons.wikimedia.org/wiki/File:Em_dipoles.svg)

density ρ_m in proportion to μ as follows (adopting the convention that ρ_m is defined in units of Webers or Volt-seconds):

$$\nabla.H = \rho_m/\mu \tag{10.8}$$

Similarly, the third Maxwell equation clearly begs for completion through the addition of a term with magnetic current density J_m parallelling the electric current density term in the fourth Maxwell equation:

$$\nabla \times E = -\mu \frac{\partial H}{\partial t} - J_m \tag{10.9}$$

There is no known physical reason why magnetic charges should not exist, which is the same as to say that there is no reason why there should be no source regions or sinks for magnetic field lines: these are so-called monopoles in the sense that known magnets are dipoles with a north and south pole for the field lines and a magnetic charge is like one of these poles in complete isolation from the other. Field lines only flow out or only flow into a magnetic monopole, as opposed to a dipole, in which the field lines loop outwards from the axis joining the two poles (Figure 10.2, lower right). The parallel with electrically charged particles and the quantum theory requires that these monopoles be particles carrying a quantized amount of magnetic charge.

Nevertheless, all observations, investigations, and experiments over the past couple of centuries, since Faraday introduced the modern concept of the absence of magnetic monopoles, have completely failed to find any trace of magnetic charges [1].

10.2   Dirac quantization

The theory of magnetic monopoles was significantly advanced by Dirac in 1931 [2]. He pointed out that, if magnetic charges exist at all, a system could be formed comprising an electric-charged particle and a magnetic-charged particle. Such a system could potentially possess angular momentum, which is necessarily quantized under quantum theory into units of $h/2\pi$, where $h = 6.62607015 \times 10^{-34}$ J/Hz is the Planck constant. This leads to a quantization condition on the product of the electric charge q_e with the magnetic charge q_m, in which N is an integer such that:

$$q_e q_m = Nh \tag{10.10}$$

The first conclusion that Dirac drew from this is that if magnetic charges exist at all, then electric charges need to be quantized (i.e. charge should exist only as integer multiples of some fixed fundamental quantity of charge). However, it is, of course, an observational fact that electric charge is indeed quantized into units of e = 1.6 × 10^{-19} Coulombs. It seems currently that the Dirac argument for charge quantization is the only known explanation for this, so it would seem that reality is at any rate well prepared to accommodate the existence of magnetic charges. This amounts to a further indication that magnetic monopoles do exist or at least can exist.

A second conclusion is that the Dirac quantization condition defines the amount of magnetic charge that would be expected to appear on a fundamental magnetic-charged particle as follows:

$$q_m = \frac{Nh}{e} = 4.14 \times 10^{-15} N \text{ Webers} \tag{10.11}$$

Dirac magnetic monopoles are singularities, but modern grand unified theories (GUTs) also tend to predict the existence of quantized magnetic charges and allow for more realistic spatial distributions of their divergence. Modern interest in the detection of magnetic monopoles is significantly motivated by their potential to provide evidence for GUTs. The actual mass of magnetic charged particles remains uncertain. There is no strict limit below the Planck mass (= 2.176 × 10^{-8} kg), so an exceptionally large mass is feasible, perhaps contributing to rarity in nature. Current experiments to detect monopoles are focusing on the mass-energy range of 20–150 GeV (3.2 × 10^{-9}–2.4 × 10^{-8} kg).

10.3   Chasing chimeric charges

Tracking down magnetic monopoles is possibly the most challenging and frustrating type of hunting ever invented by humans. There are two broad approaches to these searches. One group of hunters seeks stealthily to ensnare naturally occurring magnetic charges in traps as they roam about the countryside. The other group hires beaters to thrash the undergrowth in the hope of prompting the erstwhile concealed prey to scuttle into the open to be pursued and pounced upon by a pack of instrumentation. Despite valiant and exhaustive efforts over many decades, neither approach has yet yielded significant success, although there have been some traces of spoor.

The trappers favor a particular type of noose. If a magnetic monopole passes through a loop of wire, it induces a net pulse of current in the coil, whereas the transit of a dipole or higher-order pole induces a current pulse that integrates to zero. This distinguishes clearly between a magnetic monopole and any other induced signal and it can be monitored for long durations without attention or interruption, so such a loop is the preferred device for passively detecting passing magnetic monopoles.

In a wire of normal room-temperature metallic resistivity, the induced current is rapidly dissipated as heat energy. However, if the wire is rendered super-conducting by reducing its temperature below its transition temperature, the current pulse will persist for a significant time. In fact, the particular detector device is a variant of a basic superconducting loop called a superconducting quantum interference device (SQUID). A highly sensitive direct current SQUID is shown in Figure 10.3. The current is split into the two branches of the loop through which the magnetic flux penetrates and there is a Josephson junction in each branch. As the flux changes, it induces a current around the loop that opposes one of the branch currents and augments the other. As soon as the current in either branch reaches the critical current of its Josephson junction, a voltage is generated across the junction. In fact, the voltage output oscillates for each additional quantum of magnetic flux added to the flux threading the loop. In this way, individual quanta of magnetic flux can in principle be detected including individual magnetic monopole transits.

These kinds of searches have largely proved fruitless over many decades. Although a single candidate magnetic monopole event was recorded by Blas Cabrera Navarro on the evening of February 14, 1982 (known as the "Valentine's Day Monopole"), it proved to be a one-off and cannot therefore be reliably distinguished from some other causal activity. The total hours of monitoring combined with the dearth of detections produces a statistical upper limit on the feasible ratio of magnetic monopoles to atomic nucleons in the universe of less than one in ten to the power of 29 ($1:10^{29}$) [3]. Naturally occurring magnetic monopoles are evidently extremely rare.

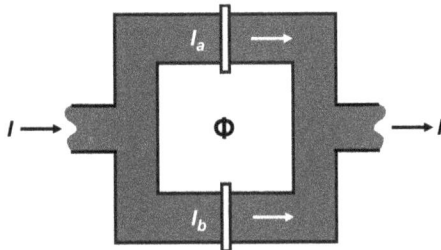

Figure 10.3   A DC-SQUID device for detecting very small changes in magnetic flux by the disturbance of the currents I_a and I_b with individual quanta of flux threading the loop and generating a voltage across the Josephson junction in either branch

10.4 Creating magnetic monopoles

The bush-beating school of magnetic monopole hunters seeks to encourage virtual magnetic charges to manifest themselves as monopole–antimonopole pairs by smashing together high-energy protons or ions to create locally extremely intense magnetic fields that should favor such production. These collisions are arranged to occur within a large particle detector array so as to obtain detailed interaction signatures that can be compared with theoretical predictions for magnetic monopole–antimonopole pairs.

This kind of approach falls within a long tradition of subatomic particle investigations at particle accelerator facilities around the world on the demonstrated principle that phantom virtual particles can be conjured into reality by using high-energy collisions between subatomic particles to provide locally intense conditions of a nature that encourage their production. The confirmation of the existence of the Higgs boson has been the greatest recent triumph for this type of research.

The most important recent experiment to look for magnetic monopoles is ongoing at CERN using the Large Hadron Collider (LHC), which straddles the Franco-Swiss border [4]. The ATLAS detector built around the particle collision zone is being used to look for the predicted secondary particle signatures from the production of monopole–antimonopole pairs. In the last decade, proton–proton collision signatures were examined in data from Run 1 (2010–12) and Run 2 (2015–18) of the LHC. No signatures for monopole–antimonopole pair production were seen.

More recently in Run 3 of the LHC during the Autumn of 2023, lead ions have been collided at 5.36 TeV energy. During these events, the ATLAS detector has been looking for signatures from "ultraperipheral" collisions, where the lead ions pass one another at an intermediate range too distant for hadron interactions to occur but close enough for intense electromagnetic field interactions with magnetic field strengths reaching 10^{16} Tesla. These conditions optimize the chances of producing magnetic monopole–antimonopole pairs, whilst minimizing the noise background from hadronic interactions. Magnetic monopoles are predicted to be extremely potent at stripping electrons off atoms, generating low-momentum electrons called delta rays, which give rise to characteristic clouds of ionization in the detector.

The Run 3 results do not appear to have seen monopole–antimonopole production. Instead, they have established a low upper limit on the cross-section for the production of magnetic monopoles in these kinds of collisions across most of the mass-energy range within which magnetic monopoles are likely to be feasible [5]. However, LHC Runs in 2024 and 2025 are scheduled to continue the search.

It might reasonably be queried as to why magnetic monopoles cannot simply be generated by chopping dipoles in two by some means. One answer is that any likely knife designed to perform such an operation would itself be fabricated from atoms or particles interacting (at long range anyway) through electromagnetic

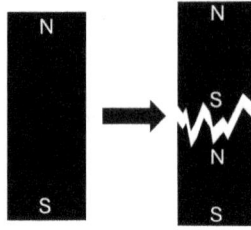

Figure 10.4   *Dividing a bar magnetic orthogonal to the axis of its magnetic dipole field generates a pair of shorter bar magnets, each exhibiting a dipole magnetic field*

forces. These fields would necessarily obey the Maxwell equations, which do not allow electric charges to sever magnetic field lines. Another answer is that in the experiment of cutting a bar magnet in a plane at right angles to its magnetic dipole field axis is that two shorter bar magnets are produced rather than a pair of isolated north- and south-charged magnetic monopoles (Figure 10.4).

It is, however, an interesting question as to whether magnetic monopoles exist at the event horizon of a black hole? In swallowing magnetic field lines within its event horizon, a black hole removes part of the field lines from the possibility of interaction with the rest of the universe, so it is as if the magnetic field has divergence at the event horizon from the point of view of our universe. Observations by the Event Horizon Telescope (EHT) team have recently confirmed that there are intense and organized magnetic fields around our galaxy's central supermassive black hole, known as Sagittarius A*. The EHT has inferred spiraling magnetic field lines from a vortex pattern of polarized light emissions surrounding Sagittarius A* [6]. Astronomers also suspect the existence of an associated astrophysical jet of high-energy ions. These jets can reach relativistic velocities, and it has been considered that tangled magnetic fields are organized to aim two diametrically opposing beams away from the central source (e.g. a black hole) by angles only several degrees wide [7]. Perhaps these jets are interesting places to explore as possible venues for future magnetic monopole hunts.

10.5   Possible technological exploitation

A starting point is the possibility that all current technology devices that rely on electric charges could potentially be reproduced in versions utilizing magnetic charges instead. However, magnetic monopoles are thought to be rather high in magnetic charge and far more energetic (tens of GeV) than electrons (0.511 MeV) or protons (938.272 MeV). These radically different properties could furnish advantages for some applications and disadvantages for others. In order to reproduce electrical and electronic devices as "magnetical" and "magnetonic" devices (assuming that we name the magnetic monopole particles "magnetons"), it would first be necessary to define and fabricate magnetic conductors, semiconductors, and insulators. This is hugely complicated and unlikely to be implemented within the next century.

More straightforward to implement would be technological applications of isolated magnetic charges. For example, pseudo-dipole magnetic fields have been used to create magnetic levitation systems driven by the strong repulsion between the non-uniform fields near like magnetic poles. Arrays of magnetic charges could similarly be used for a magnetic levitation train or other vehicle. But do such charges hold out the promise of any particular advantages over dipole magnetic fields implemented with electric current coils and/or permanent magnets?

Apart from the number of magnetic poles, another key difference between a dipole magnetic field and the field radiating from a magnetic monopole is that the dipole field tends to drop off in strength with distance rather rapidly as the inverse cube of the radius from its center. Conversely, and in common with the field of electric charges, a magnetically charged particle has a field strength that drops off as only the inverse square of the radius. Whereas this might at first appear to constitute a subtle mathematical distinction, it may have dramatic consequences for magnetic levitation technology. In current magnetic levitation systems, the levitated body needs to remain very close to the levitating field sources, whether they are electric current coils or permanent magnets. In contrast, a two-dimensional (e.g. planar) array of magnetic charges, which can be notionally horizontal, should produce a magnetic field above and below its plane that is relatively constant in magnitude with distance above and below the array until a height or depth is reached that is comparable with the radius of the array. The field would also be directed at right angles to the plane of the magnetic charge array.

A craft could be conceived carrying a like magnetic charge to that used in the array of sufficient magnitude that the repulsive force between the array and this magnetic charge would exactly balance the gravitational weight of the craft. Adjustable oriented propulsion coils driven by electric currents might then be used to develop lateral forces, vertical adjustment forces, and torques by creating small disturbances and waves in the magnetic field by analogy with the linear propulsion coils used to propel magnetic levitation trains. In this way, the craft could move about freely within the uniform magnetic charge array field, relatively unconstrained by gravity, so this would constitute a type of anti-gravity system.

A similar anti-gravity system could be implemented with an array of electric charges and a craft with a gravity-balancing like electric charge, but such an electric field would interact strongly with ordinary matter, because it contains so many electric charges. Both versions beg the question of the medium that could contain and fix a large density of like charges due to their mutual repulsion. In the electric charge version, it is feasible to charge up a sphere of conductor until the point that the surface field strength reaches the breakdown strength of air. In the magnetic charge case, some kind of cage for magnetic monopoles would be required.

Another feasible type of exploitation of magnetic charges could involve their combined use together with electric charges in future technologies. The behavior of systems comprising both electric and magnetic charges and charge flows would be expected to open up an intricate field of electromagnetic phenomena that can only be imagined at the present time. It is likely that new inventions will be required to fully exploit the technological possibilities that these new phenomena present. However, the actual specification of these new inventions is beyond the scope of the present chapter, which can but close with an invocation to future technologists to be aware of the vast opportunities that the discovery of magnetic monopoles should proffer.

10.6   Conclusions

This chapter has shown that the quest for magnetic monopoles is poised upon a knife edge. On the one hand, the indication from the Maxwell equations that symmetry in nature demands their existence has been reinforced first by Dirac's deduction that the existence of magnetic charges explains the experimentally observed quantization of electric charges, and Dirac has also defined the quantities into which magnetic charge should be quantized. Second, magnetic monopoles are predicted by modern GUTs together with indications of their likely mass range. Furthermore, devices such as SQUIDs have been designed and built that are capable of detecting magnetic charges, and particle accelerator experiments to pursue the hunt have been set up and are still underway.

Conversely, the evidence from experiments and observations to date has proved almost uniformly fruitless. The few candidate events that have been observed have had to be regarded as empirical artefacts due to a lack of reproducibility and experimental noise effects. The accumulation of evidence from these investigations leads at the very least to the conclusion that magnetic monopoles are extremely rare in nature, although there remains a possibility that they are lurking far away from Earth in exotic locations such as black holes. This possibility is enhanced by recent observations of a phenomenal magnetic field surrounding the black hole at the heart of the galaxy. Furthermore, black holes exhibit jets that harbor processes of matter acceleration involving stupendous amounts of energy, which appear to be linked to their magnetic fields in ways that are not yet fully understood.

Regarding the technological applications of magnetic monopoles, the symmetry of Maxwell's equations suggests that magnetic charges should be capable of reproducing all the useful behavior exhibited by electrons in their electrical and electronic applications. It may be that the unique properties of magnetic monopoles would make them the preferred particles for use in some existing applications. However, this also requires the availability of conductors, semiconductors, and insulators for magnetic charge currents, and it is not clear what these materials would look like or how they could be fabricated.

Uses of magnetic charges in isolation, either in oscillation or in static arrays, is more straightforward. In particular, the property of the magnetic field of magnetic charges to fall off much more slowly with range than is the case for magnetic dipoles makes them a superior source of magnetic field. For example, by deploying two-dimensional arrays of magnetic monopoles, it would be feasible to extend magnetic levitation to act over large distances, turning it into a true anti-gravity field.

References

[1] https://en.wikipedia.org/wiki/Gauss%27s_law_for_magnetism9 (accessed 8/9/2024).
[2] Dirac, P. (1931). "Quantised Singularities in the Electromagnetic Field". *Proceedings of the Royal Society A*. 133 (821): 60.
[3] https://en.wikipedia.org/wiki/Magnetic_monopole (accessed 9/9/2024).
[4] https://atlas.cern/Updates/Briefing/Monopoles-First-Run3 (accessed 11/9/2024).
[5] https://atlas.web.cern.ch/Atlas/GROUPS/PHYSICS/PAPERS/HION-2023-01/ (accessed 11/9/2024).
[6] https://eventhorizontelescope.org/blog/astronomers-unveil-strong-magnetic-fields-spiraling-edge-milky-way%E2%80%99s-central-black-hole (accessed 11/9/2034).
[7] https://en.wikipedia.org/wiki/Astrophysical_jet (accessed 11/9/2024).

Index